PAIN, SEX AND TIME

Other Books by Gerald Heard:

NARCISSUS

THE ASCENT OF HUMANITY

THE SOCIAL SUBSTANCE OF RELIGION

THE EMERGENCE OF MAN

THIS SURPRISING WORLD

THESE HURRYING YEARS

SCIENCE IN THE MAKING

THE SOURCE OF CIVILIZATION

THE THIRD MORALITY

PAIN, SEX AND TIME

A NEW OUTLOOK ON EVOLUTION AND THE FUTURE OF MAN

By
GERALD HEARD

MONKFISH BOOK PUBLISHING COMPANY
RHINEBECK, NEW YORK

Provenance Editions
Volume 3

Library of Congress Cataloging-in-Publication Data

Heard, Gerald, 1889-1971.
Pain, sex, and time : a new outlook on evolution and the future of man / by Gerald Heard.
p. cm.
Originally published: New York : Harper, 1939.
ISBN 0-9749359-1-3
1. Social evolution. 2. Intellect. 3. Consciousness. 4. Civilization--Philosophy. I. Title.
HM626.H4 2004
303.4--dc22

2004023955

Bulk purchase discounts, for education or promotional purpose, are available.
Contact the publisher for more information.

First printing

10 9 8 7 6 5 4 3 2 1

Provenance Editions is dedicated to reprints of classics of spiritual literature in elegant and durable editions. Provenance Editions are published by
Monkfish Book Publishing Company
27 Lamoree Road
Rhinebeck, New York 12572
www.monkfishpublishing.com

CONTENTS

PART III
THE DISCOVERY OF TECHNIQUES

Huston Smith's Foreword to Pain, Sex and Time

I suspect that this book would not have been brought back into print if I had not referred to it frequently as having occasioned one of the two conversions I have undergone. The first of these was my conversion from the world of practical affairs to the life of the mind. The second was when, overnight, the book in hand converted me from the scientific worldview (which takes the visible world to be the only world there is) to the vaster world of the mystics. I am in good company in owing that conversion to Gerald Heard, for Heard also converted Aldous Huxley from the cynical nihilism of his *Brave New World* to the mysticism of *The Perennial Philosophy.*

My story is briefly told. I was a graduate student at the University of California, Berkeley when the dissertation I was writing ran into the problem of pain. I hadn't given the subject much thought, so I went to the university library and checked out three books dealing with it. Back in my room after supper I found my hand gravitating to the one with the most interesting title, *Pain, Sex and Time.* It took only a few pages to realize that it had nothing to do with my dissertation, and I kept on reading. And reading. When dawn broke I was living in the new world that has housed me ever since.

Until this past week, sixty years later, I have not returned to Heard's book for fear that I would find it disappointing; we all know that what impacts us at one time may not do so later, for one can't step into the same river twice. I suspect that I might never have reread it had the invitation to write this Foreword not provided the needed excuse for interrupting other duties. I am glad it did, for the rereading has not disappointed me. It is a good book, a very good book that withstands the passage of time admirably. But before I say more about it I should say something about its author.

Today almost no one has heard of Gerald Heard, but in the second third of the 20th century the situation was otherwise; he was a well-known British polymath. He began his career as science commentator for the BBC, and H. G. Wells was widely quoted as saying that Heard was the only one he bothered to listen to on the "wireless." Heard's remarkable early trilogy of academic books included *The Ascent of Humanity* (1929), *The Social Substance of Religion* (1931), and *The Source of Civilization* (1935). *Pain, Sex and Time* came out in 1939, and in 1941 he wrote the most successful detective story of the day, *A Taste for Honey*, which sold over half a million copies, an astronomical number in those days. Thirty more books poured forth after that.

It is not the job of a Foreword to steal the thunder of a book, so I will only sketch the trajectory of this one.

Science deals with *what* happens and sidesteps *why* questions, but a complete account must include them. So with the wisdom of hindsight we can say that the object of the evolutionary process seems to have been to produce bodies with minds. When these advance to self-consciousness the mind becomes free. No longer defined by antecedent causes, it can take charge of its future.

The evolution of the human body has been completed, but it possesses two properties that suggest that its mind can continue to evolve. One of these is its inordinate sensitivity to pain, and the other is human sexuality which is not contained by periodicity and is permanently available. The surplus vitality that these properties token suggest that in humanity the life force is pressing like a jack-in-the-box for the release that mental evolution could afford. Rudiments of such evolution can be spotted in historical times in the increasing acuity of human vision and its ability to distinguish more bands in the color spectrum, but the real evolution will be in the ability of consciousness to break out of individual pockets (packets I almost wrote) and phase into God's infinite consciousness. This is more than a theoretical possibility. Mystics are the advance scouts of mankind who have transcended their egos, and in exceptional cases, merged with God completely. One more point. Methods for effecting this breakout are solidly in place. They are the spiritual techniques of yoga, meditation and prayer.

This is the vision that took me over on that fate-filled night when

I first read this book, and my pulse still quickens as I bring it to mind. But the brief account of that which I have given doesn't come close to indicating what an interesting book it is. Heard constructs his thesis from an awesome data bank, and even those who do not find the thesis convincing will find the book bristling with obscure scientific and historical facts that keep the pages turning. Some of the facts have been disproved, but enough of them remain to make the thesis still credible. If it was strong enough to persuade Huxley, I for one am still not going to dismiss it out of hand.

I applaud the decision to bring this book back into print.

December 11, 2003

WHO IS GERALD HEARD?

By Jay Michael Barrie

Henry Fitzgerald "Gerald" Heard was an author, philosopher, lecturer, and historian. His broad philosophical themes and scintillating oratory style attracted and influenced many people in England and America during the early- to mid-twentieth century. A prolific writer, Heard penned some 38 books, the most important of which are his pioneering trilogy of academic works that document the evolution of consciousness, namely *The Ascent of Humanity* (1929), *The Social Substance of Religion* (1931), and *The Source of Civilization* (1935).

Gerald Heard was born in London on October 6, 1889, the youngest of three boys. His family home was in Ireland, where the Heards have been landed gentry since the 16th century. Having finished public school at Sherborne School in Dorset, his university years were spent at Gonville and Caius College at the University of Cambridge, where, in 1911, he took honors in history. He remained at Cambridge on a scholarship to study theology in preparation for a career as an Anglican clergyman. He subsequently never pursued ordination, because in 1916 he suffered a crisis of faith and became, for some years, a secular humanist. During that time he was involved with progressive education, prison visiting, and social reform.

By December 1919, Heard began working as a secretary to Sir Horace Plunkett, founder of the Irish agricultural cooperative movement, living first in Dublin and later in England. While in Ireland, Heard came to know many of the notables of the time, including George Bernard Shaw, W.B. Yeats, Lord Fingall, George Russell, Colonel E.M. House, and Lady Gregory. Heard penned his first book, 1924's *Narcissus: An Anatomy of Clothes*, while at Plunkett's home in England. The thesis of this book is that fashion through the ages provides clues to the evolutionary stages of mankind. Many of the ideas set forth in *Narcissus* found expression in his subsequent work.

Heard began his career as a public speaker in 1926, lecturing for three years under the auspices of Oxford University, and he began making a name for himself as a scientific journalist. In 1929 he was brought in as literary editor of the short-lived monthly journal, "The Realist." There he worked with an editorial board that included Julian and Aldous Huxley, Arnold Bennett, Rebecca West, and H.G. Wells.

In 1929 Heard published his second book, *The Ascent of Humanity*, an essay on the philosophy of history that contested the conventional drum-and-trumpet interpretation of history that only records the "crimes, follies, and misfortunes of mankind" (Edward Gibbon). History, he proposed, is "the shadow cast by the evolving consciousness of man." Humans can—and must—accelerate their own evolutionary development by advancing in consciousness if they are to survive. *The Ascent of Humanity* was awarded the distinguished Henrietta Hertz Prize by The British Academy.

From 1930 to 1934, Heard served as the first science commentator for the BBC, commanding a large and regular listening audience with his fortnightly broadcasts, during which time he had instant access to all the greats in the world of science, both in England and Europe. As H.G. Wells once said of him, "Heard is the only man I ever listen to on the wireless. He makes human life come alive." Throughout this period he acquired a working knowledge of general principles in the various scientific disciplines that, coupled with his encyclopedic retention of facts, would furnish him with many apt illustrations for his writing and lecturing in later years. His circle of acquaintances at this time included J.R. Ackerley, W.H. Auden, G. Lowes Dickinson, E.M. Forster, Naomi Mitchison, and Harold Nicholson.

By the early 1930s, Heard had become familiar with Hindu and Buddhist thought. Inspired by the Buddhist proposition of following a path to enlightenment stripped of theology, he transitioned from secular humanist to practicing mystic as he undertook meditation and related spiritual practices. His 1937 book, *The Third Morality*, explores this theme by documenting three historical Western worldviews and the moralities that are sanctioned by them. Complementing these sections is the psychophysical training or "athleticism" that individuals must undertake in order to collectively advance society toward spiritual salvation. These exercises involve regulating diet, attitude, respiration, and meditation.

In April 1937, together with his close friend Aldous Huxley, Heard left England, never to return, and arrived in New York City. He had been offered the post of Chairman of Historical Anthropology at Duke University, but he decided to leave that position after delivering a series of lectures in that capacity for one term, feeling that university life would be too confining. He and fellow pacifist Huxley, both long associated with England's peace movement, embarked on a cross-country joint-lecture tour on world peace. After exploring various sections of the country, Heard settled in Southern California, which was to be his home until his passing in 1971.

In 1939, Heard met Swami Prabhavananda, founder of the Vedanta Society of Southern California, and began, under his guidance, the study and practice of Vedanta. Although this ancient Hindu philosophy gave him his final philosophical frame of reference, his curiosity-ridden mind never ceased exploring many diverse areas of interest throughout the remainder of his life. His influence drew in Aldous Huxley, Christopher Isherwood, John Van Druten, and others to the Vedanta circle. Heard co-edited the Vedanta Society's journal from 1939 to 1941 and contributed many articles to it over the years. He later served as an editorial advisor to the journal from 1951 to 1962.

A long-cherished dream of Heard's had been to establish a place where the study of comparative religion, together with research into and practice of the techniques of meditation and prayer as taught by the major religions of the world, might be carried on. In 1941, Heard put the larger part of his personal financial resources from his inheritance into building and endowing Trabuco College, which was completed in 1942. It consisted of a large complex of Mediterranean-style buildings situated in the middle of some 300 acres, about 75 miles south of Los Angeles near the then-remote and small community of Trabuco Canyon.

For the next five years, from one to two dozen revolving students meditated three times daily, studied, performed necessary chores, worked in the garden, prepared and ate vegetarian meals, and listened to lectures on religious life by Heard, Huxley, and others. Heard wrote several books there, while Huxley is said to have written *The Perennial Philosophy* while at Trabuco. However, because of a lack of permanent residents that were needed to maintain the buildings and immediate grounds, the Trabuco College experiment was finally discontinued in 1947. In 1949 Heard turned over the property to the Vedanta Society of Southern California, which subsequently named it the Ramakrishna Monastery.

The 1940s, Heard's most productive decade of writing, saw no fewer than 18 books published. His creative output, seemingly inexhaustible, resulted in essays, articles, short stories, introductions, and epilogues appearing in a variety of publications. In addition to his non-fiction writing, Heard authored several mysteries and supernatural fantasies under the pen name H.F. Heard, including *Reply Paid*, *The Notched Hairpin*, and *Doppelgangers.* His 1941 novel *A Taste for Honey*, praised by Christopher Morley and Boris Karloff, was loosely adapted into a movie, 1966's *The Deadly Bees*, the first in the killer-bees genre. His 1947 whodunit, "The President of the United States, Detective" won first prize in the second-annual *Ellery Queen's Mystery Magazine* short-story contest. Translations of Heard's works have appeared in French, German, Japanese, and Spanish-language versions.

For the remaining fifteen years of his active life, Heard spent his time and energy in writing, lecturing, research, travel, and making numerous radio and television appearances. His bearded face, ready wit, and sonorous voice were a familiar sight and sound on television screens in the 1950s as he addressed himself to such diverse subjects as history, literature, and mental health. He moderated an eight-part series, "Focus on Sanity", which appeared on CBS in 1957.

From shortly after his arrival in America to the early 1960s Heard lectured at most of the major colleges and universities in the United States, including Harvard, Cornell, Princeton, UCLA, UC Berkeley,

USC, Mills College, Rockford, and Wabash. At Colgate Rochester Divinity School he delivered the Ayer Foundation Lectures that would become 1946's *The Eternal Gospel.* At the Oberlin Graduate School of Theology he was the Haskell Foundation lecturer in 1958. He spoke at Christian, Jewish, Hindu, and Buddhist religious venues. About 300 recordings exist of his lectures during these fruitful years.

Heard spent the better part of 1951 and 1952 in St. Louis as visiting lecturer in philosophy at Washington University on a fellowship from the Danforth Foundation. Out of these lectures came his penultimate major work, *The Human Venture.* Heard received a two-year fellowship grant from the Bollingen Foundation (1955-1956) under the auspices of Washington University's Department of Philosophy, then headed by Professor Huston Smith. This enabled him to undertake the research that resulted in 1963's *The Five Ages of Man*, of which Robert R. Kirsch, then literary critic of the *Los Angeles Times*, reviewed as, "...the most important work to date of this challenging and brilliant philosopher, a volume which in scope and daring might be the *Novum Organum* of the twentieth century."

In the late 1950s and early 1960s, Heard made six narrative record albums, including "Re-birth", a contemporary interpretation of *The Tibetan Book of the Dead*, which provides rites of passage for the dying. Despite leading such a busy life, he maintained a regular schedule of meditation, lasting six hours daily for many years. Heard wanted nothing more than to attain spiritual liberation. Although his personal lifestyle was abstemious to the point of being monastic—he was celibate by choice for the latter decades of his life—his rule of thumb for moving about in the world of people and things without being offensive or a nuisance echoed the advice of his favorite saint, François de Sales, "Ask for nothing, refuse nothing."

Gerald Heard was a pioneer and a catalyst; a bold, original thinker who was ahead of his time. His groundbreaking early philosophical works interpreted history in the context of mankind's evolving consciousness. His innovative cosmological schema depicted mankind's evolutionary purpose as fundamentally spiritual and teleologically oriented; it incorporated data from disciplines as seemingly diverse as anthropology, archeology, astronomy, biology, ethics, history, mysticism, mythology, paleontology, physics, physiology, psychology, religion, and sociology into a unified whole.

An inexhaustible researcher, Heard investigated numerous areas of human activity. For ten years, from 1932 to 1942, Heard was active on the council and the research committee of the Society for Psychical Research. His 1950 book, *Is Another World Watching? The Riddle of the Flying Saucers*, among the first full-length works on UFOs, was completed in three weeks. In the early 1950s he helped pioneer investigations into the value of psychedelics in accelerating spiritual growth and social reformation. His ideas on sexuality, viewed as a force that could be harnessed for spiritual evolution as outlined in *Pain, Sex and Time*, and his theories on homosexuality as an evolutionary, spiritual phenomenon, were maverick. Trabuco College was the first coeducational, spiritual community in America to incorporate ecumenical, non-sectarian religious principles and practices.

Preferring anonymity to the limelight, Heard was a behind-the-scenes inspiration who spurred many individuals to pivotal accomplishments in their particular field of endeavor. His circle of friends included Henry and Clare Luce, Igor and Vera Stravinsky, Steve Allen, Edwin Hubble, Dave Brubeck, Nancy Wilson Ross, Mary Wigman, Malcolm Sargent, Fr. John Courtney Murray, E.M. Forster, Dr. Sidney Cohen, Morris Graves, Ethel Barrymore, Aldous Huxley, Roger Fry, John Gielgud, David Kahn, John Van Druten, Kenneth Clark, and John Betjeman.

Gerald Heard once remarked, "If I am remembered at all, I hope it would be as an historian of consciousness and its evolution." Heard held the conviction that history traces humankind's progressive stages of psychological and spiritual growth. Armed with a spiritual worldview—the core beliefs that a person espouses about the universe and themselves, and the frame of reference by which they interpret and understand life—and an ethical lifestyle that supports it, Heard maintained that a person is able to self-consciously accelerate the evolution of their own consciousness. This can be accomplished by employing various psycho-spiritual techniques such as meditation.

On February 5, 1966, Gerald Heard suffered the first in a series of strokes that left him increasingly incapacitated. After several years of a slow decline, Heard succumbed peacefully on August 14, 1971 at his home in Santa Monica, California, at the age of 81. The gallantry and courage with which he faced an increasingly cruel ordeal during these years was nothing short of heroic. His many years of inner discipline gave him not only fortitude but serenity and a cheerful acceptance.

****

Jay Michael Barrie (1912-2001) was Gerald Heard's personal secretary and editor, associated with Heard from their meeting in December 1944 until Heard's death in August 1971. John Roger Barrie is the current literary executor of Gerald Heard.

For more information on Gerald Heard, visit www.geraldheard.com, the Official Gerald Heard Website.

INTRODUCTION

"SHORT views, for God's sake, short views," cried shrewd, kindly Sydney Smith. His generation, reared in the classical assurance of unchanging order, found itself at middle life in the tornado of Revolution. The Romantics who welcomed the sweeping away of the stately porticoes in which their elders had strolled, delighted in the thought of the vast deserts, the magnificent desolate prospects which were being revealed, as arcade, façade and artificial vista fell in ruin. Their fatal exultation could not be checked. It was in vain that the Conservatives demonstrated that the Romantics were deluded, that man could not live at one with nature, that you had to have the mediation of form and order and that an accustomed good form and a long acquired order were easier and better than a dictatorship. "The Conservatives are always right: the Progressives always get their way."

Hence when it was clear that the advanced people were going to succeed, would liquidate the eighteenth century's notion of a classical culture open to all the educated and transcending political frontiers, and put in its stead perfervid, mutually exclusive nationalisms, then long views changed from being a solace to being an added misery. The philosophically minded deliberately retracted their vision.

Our present situation parallels Sydney Smith's, but on a lower level. He saw the beginning of a degenerative process of which we are witnessing the end. The international anarchy, which in his day increased from the stage of occasional professional duels to the raising of whole nations *en masse* against one another, and involving the educated, the savant and the philosopher as much as the common people, has now reached a final phase.[1]

The last vestige of the belief that those who have the same material techniques and powers have the same moral standards and aims, has vanished. Hence Sydney Smith's cry finds more than an echo in most educated minds today. Degeneration seems to them certain.

Their information but makes more clear to them the descent which civilization is following down to the Pit. They feel they must learn somehow to do intentionally what the simple do without thought and, though to-morrow everything be in ruins, avert their eyes, concentrate on the present and enjoy to-day.

This is inevitable. In a local and brief disaster long views help. Though our day is spoilt, other unspoilt days will come, if not for us, for others. In a disaster which involves civilization and indeed humanity, long views, the vision of reverse upon reverse, the collapse of defence after defence, the foundering of land after land under the Deluge, defeat the last remnants of the will to live.

It is that discouragement among thinkers which is the gravest symptom in the present international Depression. The men who should through their vision keep serene, content and perseverant, are, through what they foresee, more daunted and paralyzed than the unreflective. In brief we are in a peculiarly serious crisis because belief has broken down. Yet much of our discouragement rises from an inaccurate diagnosis. There is, no doubt, reason for gravity and even apprehension. But we gain nothing by misapprehending the nature of the disease with which civilization is attacked. Indeed, when we have made an accurate diagnosis a cure may appear, and though it may be drastic and lengthy, it may at least relieve us from having to cry "short views" and from having to try, what is probably impossible for the reflective, to become unreflective.

The first grave misapprehension that we make is that our civilization's crisis is a repetition of the last. We talk of the Decline of the West and think of the end of the Roman Empire. The cases are not the same. The Roman Empire died of a social pernicious anæmia. Simultaneously all its forces waned. Its wealth declined as did its health: its population as did its revenues: its invention, resource and

---

1. It is worth reminding ourselves as an indication of the fall in the moral barometer since that time, that during the last great ideological crusade—the Napoleonic Wars—Sir Humphry Davy, the English scientist, although a discourteous and aggressively patriotic man, is asked to Paris by the French savants whose government is at war with his, and is received with hospitality.

skills as did its bureaucratic efficiency, its official integrity and its military discipline. The waning belief in the Empire's right, its immense meaning and its cosmic *raison d'etre* was accompanied by an equal decline in the quality of service it gave to its subjects, an equal increase in the burden which it imposed upon them, and an equal assurance on their part that the significance of life lay beyond its authority, its protection or its influence.

Such is in no wise a description of our social disease, acute though it be. If the Empire died of pernicious anæmia we are attacked not by that but by convulsions, by a social epilepsy. Its means were inadequate to its vast purposes. With us the reverse is true. We need not, and therefore must not expect exhaustion and that, becoming empty, we shall be overrun as were the depopulated areas of the Empire. What we should have foreseen, are experiencing, and must if possible find a cure for, are convulsions. That is not to say that our outlook is not serious. Indeed, it is more immediately acute than any crisis the Roman Empire endured. Rome died, and could only die, by inches, its decline so gradual that men still dispute when it actually fell; those living at its dissolution hardly realized it was dead and had never known it as really alive. Our civilization could collapse all at once, in a single convulsion. The interramifying workings of its structure, economic and psychological, penetrate below and across the military frontiers and those passages are all mined, until one detonation might explode its foundations.

This is due to no peculiar or sudden perversity on our part. On the contrary, it is a result not of any change in us but of our refusal to change. Without thinking of what we were doing, without realizing that if you keep on increasing the load on the port side of a ship you must put an equal weight on the starboard, we have incessantly and industriously increased our means, our physical powers, without doing anything to advance or even to elucidate our aims and ends. If aims and values are not kept advancing in clarity and comprehensiveness as fast as powers increase, that can only mean that the train is pushing the engine off the track and all will be wrecked.

We cannot escape this conclusion. An adequate end, a goal worthy of a man's wholehearted and lifelong endeavour has, *pari passu,* faded out of life as means increased. When, in the Middle Ages, means were moderate and stabilized, the goal generally accepted as Life's meaning was immense. Man lived A. D. M. G.—to fulfil the purpose of the universe's Creator and to be rewarded eternally if and when he so succeeded, being punished eternally if and when he failed. This life of some thousand months, at a handsome estimate, settled his fate forever and that fate was incomparably important.

Then came the beginning of the Modern Age with its inventions leading at once to an increase of powers, and precisely at that time the goal of life shrank proportionately. This life became the end of all but exceptional men. The belief in continual invention leading to a continual increase of means, of "goods," made it possible for men to believe that this individual physical life might in itself become a sufficient end, and that all men's questions about eternity and desire for a heaven might be dissolved when everyone had complete physical satisfaction. The advances in astronomy at the same time made this supposition the more welcome, as they seemed to prove that there was no heaven in which the abstinent and the spiritually minded could be rewarded, or the indulgent and wicked judged. A further great increase of powers was followed by a further attenuation of ends. The goal of physical satisfaction would not, it was now realized, be attainable in this generation, nor in the next. Indeed, no one could say when. But there was working in man a natural evolution which led to inevitable progress. Inevitably, if at some unknown date, man could and must attain complete physical satisfaction, complete happiness in the body, if no cosmic cataclysm destroyed him before then and if, when he had attained, the attainment should prove such a soporific that he would have no misgiving about the future. For the whole of humanity must finally be exterminated by the earth becoming uninhabitable. In our own time an immense acceleration of invention and increase of means again led to a proportionate shrinkage of ends. Inevitable Progress and the goal of earthly happiness for all, followed into the discard the heavenly goal for the good and the saved. Its place was taken by an even less satisfying objective. By Revolution a few right-minded people would destroy the wicked, and the oppressed would enter into Utopia.

True, the goal was brought much nearer than it had been in the phase of Inevitable Progress. But it was restricted to a class and it was to be won at a price. As the years of revolution have gone on, the date of the victory has had to be postponed, the cost increased and the number of the elect reduced.

Finally in this generation the psychological revolution (following the religious, the political and the economic) has, as each successive one did to its predecessor, "debunked" the earlier crusading faith and left its followers without zeal and bewildered. For the individual now realizes that it is the conflict in himself that he projects on the world without, and which causes the confusion in which he suffers there. Each protestant, each republican, each proletarian carries the seed of schism, secession and class consciousness ineradicably in his own sundered psyche.

This doctrine now appears as undisputable as it is despairing. Nor is it any longer merely a psychological theory depending for its proof on private practice among exceptional neurotics. We have demonstrations on the largest scale. We see that whereas the few individuals suffer from individual neuroses, the great masses are now reacting to the same condition by mass neurosis. The so-called normal individual causes Totalitarianism because he cannot live without an end adequate to control and to express his means. His individualism cannot be that, is totally inadequate for that, hence he strives to create out of his nation a being in which he may experience a preindividual solidarity.

Unless, then, we can find an end really adequate to our means, a true meaning and purpose of life as an entirety, the only choice before us now is either individual neurosis or mass neurosis. We shall retreat into solitary fantasy, or into group fantasy, we shall decline to carry on at all or we shall carry on but as a series of criminal conspiracies against Life and against each other, ending in our destruction.

But what end, adequate to our present means, can we find? The reaction of the individual shows clearly what his psyche demands. By losing his adequate goal, in the pursuit of ever more powers, he has become neurotic. He must find some psychological state in which he escapes from the torturing frustration which is experienced by a self-conscious individual faced with a universe which gives his consciousness no meaning, which must render all his activities sooner or later futile. But where can he find such a state which is not either solitary fantasy or a fantasy of solidarity? Can he go on, or must he, even for a temporary if fatal adjustment of his distress, go back to a mythical past? It is at least clear that as Individualism is a neurosis and that man, attempting to cure that unbearable condition, makes his life and civilization only worse by attempting to escape into a preindividual condition, the only real cure would be by emergence on to a post- or supra-individual state. Nor would this only assuage his distress of mind and body without destroying his estate. It would also, and it alone, give at last an adequate end to match, control, direct and express his immense means.

Such a supposition would, naturally, never occur to any of those revolutionaries who seek to cure man's inner distress by altering his outer circumstances, who would relieve his misery at the disappearance of satisfying ends by increasing still further his means (although the increase of means has, as we have seen, been the invariable concomitant throughout the Modern Age, of a diminishment of ends). Nor does this way out of our dilemma occur to the psychologist nor to the religious, any more than to the economist or the materialist. The psychologist sees no future for man. His one aim is to make his patient normal, to return him to that condition of the present which, it is now obvious, cannot last. Indeed, between the public, actual world, as mass neurosis now shapes it, and the delusional, private world as the neurotic images it, it would be hard to say which better accords with truth, which is a more practical reaction to reality; while it grows increasingly evident that the latter is the more satisfying, as the pain of a broken back is more endurable if one is drugged than with clear consciousness.

The religious also see no future for man. Muscular Christianity, Utopian Unitarianism are both doing badly to-day. The Churches which are lasting are those which have the courage to stick to other worldliness—at least if they are fooled in the end, they will be fooled so much the last that none of their rivals will be able to exploit their discomfiture.

Yet this possibility of evolution continuing should not be wholly preposterous to modern mankind. We are now living in the third generation since the idea of inherent development became part of educated mankind's basic thinking. If we are at a point where only a change in human nature would extricate us, is that a wholly impossible notion for us to entertain? Of course that is to carry on the idea of development from the physical to the psychical. But again, is that to be dismissed out of hand? No one expects evolution by physical change, even were that method of development still working (which we shall see there is reason to suppose it is not), to take place in time to make any difference to human civilization. The life of a civilization is some thousands of years: the time taken for significant structural changes to take place in a species (even though such changes may be built up by instantaneous mutations) is certainly hundred of thousands of years.[2] Bodily Evolution has, we may say with considerable certainty, ceased with mankind, but we can say with exactly the same amount of certainty that it was when man's physical evolution ceased that his mental evolution began and the latter may quite likely have been the reason for the former. As we shall see, in the succeeding chapters of this book, all man's specific history may be interpreted as the symptoms of a mental evolution. There lies, then, a case for examination. It may be that if man's civilization is the shadow cast by his evolving consciousness, then that evolution of the psyche is a sequel to the evolution of the physique which could itself go no further. In those circumstances human nature, having changed and evolved by inherent laws of growth, could again change, and so man, by a mutation, this time in his psyche, and not in his physique, might rise to his present crisis. He might evolve another species, not different in physique and organs but in psyche, in consciousness and faculties. He might attain to another and further state and stage of being.

This is not an impossible hypothesis, it must be repeated. It is, moreover, the only one which offers us any adequate hope. Nothing less will extricate us now. It remains to see whether the facts fully

2. See Osborn on the Development of the Titanotheres.

sustain it. Is it possible to postulate that evolution in man might now take this steep, sharp angled turn in the spiral of ascent?

Let us first see the common assumptions which to-day informed opinion generally makes about evolution. Three postulates are now generally accepted about man. The first is that he is part of an evolutionary process which has gone on since the dawn of life. The second, that he is a culmination of that process. Never before has a single species spread over the entire land surface of the earth, dominated completely all other species, radically altered its environment by tools, instruments and machines and, passing beyond simple awareness, attained self-consciousness. The third postulate is that evolution progressively accelerates until, in man, it has reached its highest speed. This acceleration is so rapid that man, the final species, in his latest stages passes in a millennium through changes comparable to those which in the dawn of life took a score million years.

There are two further postulates about which there is growing agreement among informed inquirers. The first is that evolution has ceased in all other species; in man alone is further radical change possible. All the others have specialized: man is still unspecialized. The second of these postulates is that though man is, through his retention of an unspecialized condition, capable of further physical evolution, he does not seem to have availed himself of this, his latent capacity. On the contrary, he appears to have kept himself "in solution" for an unprecedented time: indeed it might be said that he has, as far as bodily development is concerned, specialized in unspecialization.

From these postulates it would seem that we are compelled to go further and to advance an hypothesis. Granted that physical evolution seems to have ceased in man, and yet he, and he alone, retains the capacity for further advance, where should we look for that advance? The hypothesis put forward here is that man's development when it ceased to be physiological became psychological. His history is his specific evolution. He evolves mentally. Even if we had no record of what this creature has done since its physical evolution ceased, we might assume that it would be its intelligence which would expand. We should expect to find that this creature had produced a culture, "projecting" on to its environment the energy which would otherwise have been expressed through further alterations in the creature's physique.

This is to bring history under the general principles of natural history, to make civilization a development of biology, and to see culture as a continuation of a vital energy which, no longer manifesting itself in changes in physique, begins to express itself in discoveries of technique. This principle also allows us to diagnose not merely the growth of the technique phase, as an extension of the evolutionary force which could go no further in physique and had, if it was to continue, to find another method of advance. It also permits us to understand and make a prognosis of the technique phase, of the real nature and possible future of civilization. We see that as change in physique came to an end, man, if he was to continue and not become like all other animals, either a prey to morbid outgrowths (hypertrophy) or arrested, had to produce a completely new method of evolving: the technique way. So, in turn, this technique way has done for him all that it can. It has brought him, as his physical evolution brought him, to a stage where he must leap forward or sink (or collapse) back. The signs of decadence in this technique phase of evolution are as clear as they are in the physique phase, and similar. The symptoms that a species is becoming decadent are, in the main, two—the multiplication of specializations, regardless of a general coordinating purpose, and the increase of means, the accumulation of resources and tissues, while the ends, for which such resources could be used, become increasingly faint and finally obscured. These two symptoms, as we have seen, are now showing clearly in the technique phase of man's evolution, his civilization. He is, then, at an end of that second of his lines, or traverses, of advance, as three or four-million years ago he was at an end of the line of simple physical advance. He must either hypertrophy and his common coordinating civilization disappear, consumed by a series of isolated malignant growths (the anarchic nations), or he must find still another departure or mutation, as radical as was technique progress from physical progress.

Nor need we be in doubt as to the direction of this next departure.

The first great stage of advance was the physical, the second was the technical, the third must be the psychical. The first is unconscious—blind; the second is conscious, unreflective, aware of its need but not of itself, of how, not why; the third is interconscious, reflective, knowing not merely how to satisfy its needs but what they mean and what the Whole means.

We must, however, remain completely ignorant as to whether man will do this, for at this stage unconscious evolution definitely ends. Further advance, though natural, though possible, though endowed, must be wholly intentional. The first vast phase of advance, the physical, was unconscious, and though it did arrive at the production of a conscious creature out of inorganic material, it followed so long and, some think, so hesitant a path that they would deny that it had any purpose. The second phase divides into two stages—the span of the Paleolithic cultures and the short terminating section of actual civilizations. As the Paleolithic cultures are much shorter than the evolutionary æons, they also show a proportionate increase of intention—purposive tools are made, if almost as unconsciously as a new organ is evolved. And as civilization's inventions grow clearly conscious, so the time required for radical technical advance becomes proportionately shortened.

When, then, we again change, this time from the technique phase of advance to the psychic, from indirect to direct expansion of understanding, at this point man's own self-consciousness decides and can alone decide whether he will mutate, and the mutation is instantaneous.

Such a view of history, by bringing it into line with natural history, has nothing in it that is revolutionary. Rather it should be assuring to all naturalists and especially to our generation, which has viewed with general pleasure (one of the few which all civilized men can now enjoy) the advance of those Principles of Relativity as they brought ever larger ranges of physical reality under simple and comprehensive generalizations.

Nevertheless this hypothesis is unfamiliar and unacceptable. Having abandoned the belief in simple physical progress, biologists and historians are shy of any further hope. Having decided that, "It is only on a foundation of unyielding despair that the soul's habitation may be built," they are unwilling to shift from that cool, sequestered seat back into the light of any meaning or rational philosophy, especially when that philosophy is one of action and demands wholehearted effort. "Wishful thinking" can be as easily negative as positive. Having adapted myself to senselessness it is as painful once again to have to accept meaning as, long ago, it was to abandon it. I shall resist as actively arguments which would convince me again that I must cooperate, as once I resisted the arguments which showed that my cooperation was senseless.

To tell men there is hope when they have abandoned it, is as painful to them as to take it away when they are still clinging to it. It is as agonizing (indeed often more so) to regain as to lose circulation by frostbite, to regain breathing and consciousness from drowning as to lose it. Our age is one of growing discouragement. It must be so. We have seen why. As Proust remarked more than twenty years ago, "We say spontaneously, 'Too good to be true,' never 'Too bad...'"

A hopeful hypothesis must then be supported with evidence which would not have to be maintained by one which chimed with the tolling of the current curfews.

The hypothesis of this essay will, therefore, require the following somewhat detailed substantiations, if it is to receive any attention. Those who find such confirmatory particulars tiresome must blame the discouraged spirit of our time. It takes considerable time and much patience (when time is short and patience scarce) to convince a horse in a burning stable (a simile to which we must revert at the end, for it is sadly apt to our condition) to leave its smoking stall and come out into safety.

The first chapter gives the evolutionary evidence which indicates that in the entire advance, from the most primitive forms of life up to the completion of man's physique, the one clear coordinating achievement is heightened awareness. The task of life is the retention of general awareness and the avoidance of any partial apprehension or absorption which would blunt or restrict that expanding sensitiveness. In following this master clue through the jungle of geology, Dr. Robert Broom's *The Coming of Man* proves an invaluable guide. It is a book which no modern historian or natural historian can disregard. For the final period, of man's specific rise, Professor Le Gros Clark's work, *Early Forerunners of Man,* is important, giving as it does the physical evidence that in man, as in all the previous forms which give rise to his stock, the same principle, of unspecialized awareness, was decisive.

The second and third chapters assemble the evidence for believing that in man and in him alone is now left a store of evolutionary energy and that that energy can give rise to his further, purely psychical evolution, but that, balked as it now is, its only adequate outlet is that which is threatening and will destroy civilization and humanity.

The fourth chapter attempts to indicate how man's evolution is not only continuing but that there is no break in this stream of evolution, which raised his stock from the level of the lowest creature to where it is without animal equal. We are not compelled to believe that evolution must flash over from the bodily to the mental, from physique to consciousness, without any intermediate stage or step. There is remarkable evidence that man does go through a middle and bridging stage of evolution, half-way between purely physical and purely psychical evolution, between development through change of physique to development through change in consciousness. This stage is illustrated by the evidence for the growth of man's senses, especially the emerging dominance of sight over all the rest, in particular over smell, and next, sight's acquisition and expansion of a wide-ranged colour awareness.

The fifth, sixth and seventh chapters will trace in an historical outline with an attempt at chronological correlation, the emergence of the specifically intentional evolution of consciousness. These chapters will attempt to show that it is only in so far as man can intuitively or intentionally balance the growth of his mind, and understand himself as well as he understands his environment, that he can continue evolving and not relapse into strangulated self-consciousness which gives him means without ends and powers without sanctions. The meaning of history is here seen as the attempt on the part of historical man to co-operate with his rightful evolution, to gain increasing intensity of understanding without losing or contracting his general awareness; to retain into full consciousness his profound apprehension of the meaning of the whole, of every part's unity in that whole and of his vital co-operation with them and with It.

The concluding chapters will therefore deal with the last epoch—in the last phase of which we are living—when the increasing intensity of awareness no longer permitted man to retain his sense of the whole (and of the laws and sanctions such a whole imposes on its parts). We shall see that he did not at once act on this warning. He did not instantly set about bringing his psychology, his knowledge of his psyche, up to the same level of intense knowledge which his physics had reached. Finding his intuitive psychology (at this stage rendered in the degenerative forms of anthropomorphic religion) hopelessly out of date and in conflict with his physics, he simply let his religion go and with it, inevitably, after a while collapsed his ethics, of which religion had been the inadequate but only sanction. Now, however, with his sanctionless ethics in ruins, and no longer capable of restraining in any wise his physics, he is compelled to think of sanctions. He becomes aware that he must understand his own nature, his psyche. He realizes that vague and ever vaguer intuitions in psychology and ethics will not be able to balance, control and direct clear and ever clearer knowledge in physics. He must make a science of himself, and have a power over himself, as reliable and as effective as the science of his environment.

He must discover why he experiences this paralysing and destructive conflict. He must go back to where it became acute and there learn how, when it was mild, by what methods it was kept under control. From those old methods he may learn how to devise new ways enabling him to master his present conflict. In short, he must relink himself, by self-knowledge, once more to his evolution, find where this interior force became sundered from his consciousness, repressed, balked and therefore deadly, and by giving it intentional outlet, by understanding its purpose in and for him and co-operating with it, save himself and let evolution resume.

Yet this is much to ask of himself. For it will appear that it is his individuality which causes the trouble. It is the cause and symptom of his sundered and thwarted psyche. Nor is that all. To assuage its misery it has created an environment—the present competitive, internecine, mechanized, militarized world—which prevents its cure, its reduction.

If, then, in us and through us evolution is to be resumed and to continue on to new levels (new levels of consciousness) it will need something more than any private, subjective resolution. The continuation of evolution consciously to higher consciousness needs complete devotion, complete knowledge and a complete way of life, so as to give the psyche those conditions under which all its efforts at growth will not be thwarted by its circumstances. As well hope to continue evolution in our competitive society where the aims are to satisfy the ego and (instead of eliminating it) to stabilize it by addictions, possessions and pretensions, as to rear, from seed, orchids in the arctic.

The concluding chapters therefore give indications of the psychiatry, the economy and the policy which are the minimum if man is to attain that trained condition in which alone his further evolution is possible.

PART I

EVOLUTION THROUGH THE PHYSIQUE

CHAPTER I

THE SIGNIFICANCE OF PHYSICAL EVOLUTION

The Task of Life Is to Free Consciousness from Accident

WHEN we see Evolution as a whole it divides into three fundamental stages. The first stage is development through physique. It has taken, according to modern computation (through testing the radioactive content of fossil bearing rocks), some fifteen hundred million years. In that span of time the psycho-physical organism passes through five epochs, during each of which it attains increasing intention—power to direct itself and ability to avoid accident. Beginning with a single-celled creature, its next step is to pass to a collection of cells (such as Volvox) which hang together but which, beyond this capacity have hardly any other joint achievement or specific differentiation.

The next epoch is marked by the appearance of directional organisms, bodies which have point and drive toward certain objectives. This epoch leads to a differentiated organism possessed of specialized members. It begins with such early creatures as the sea scorpions. Its middle and most enduring form is represented by the fishes and its final term is the Saurians, the giant lizards which disappeared between one hundred and fifty and one hundred million years ago. In this term of development, although the creature is possessed of specialized members, considerable "dry-land mobility"; they no longer crawl but stride (Iguanodon), run (Struthosaurus), and even fly (Pterodactyl); and the concentrated orientation given by a neck-extended head, nevertheless, the brain is not dominant. It is not even *primus inter pares* among the great nerve centres. In Diplodocus it is apparent, from the cavities in the spinal column, cavities considerably more capacious than the cranial cavity occupied by the true brain, that in the spine, between the forelimbs, and in the lumbar region were great ganglia (the lower no doubt being similar to our solar plexus). These nerve centres it seems, must have run their segments of the giant body without much consultation with the rudimentary brain. There cannot therefore have been a single continuous conscious purposive activity in the creature, which probably functioned through a series of reflexes. Dr. Swinton of the British Natural History Museum—an authority on the Saurians—thinks that the lack of intelligence must have been so complete that even with the active Carnivorus Saurians, such as Tyrannosaurus Rex, stalking or lying in wait for prey must have been an impossible feat of intelligence.

The fourth epoch begins with the mammals, when the brain has won the position of being at least *Primus inter pares.* True consciousness becomes possible and, though it is not yet continuous, when it is not directing, then the organism waits, outer activity is in abeyance, "the House stands adjourned." No secondary center can behave like a secessionist and carry on wholly independent activities. As, in the political evolution to centralized government in France during the seventeenth century, the various great provincial lords were brought to court and had to live under the sovereign's eye, so now the levels of secondary consciousness become centralized. The body-mind changes from a widely distributed feudalism to a centralized bureaucracy housed in the cranium.

The fifth epoch is that of man. The brain is now the seat of a self-conscious mind and that mind is capable of interests and aims which are wholly beyond the body's grasp and satisfaction. The mind which was a means to the body has now begun to reduce the body to a means of the mind.

These five epochal steps in evolution are possible, and only possible, because enough general energy and future aiming drive are retained in each developmental phase through which the organism, the bodily form, goes, from unicellular organism to man. This energy drive is sufficient to raise the creature to each successively more conscious, more intentional and more widely ranged power, because at each epoch the creature retained the two physical characteristics without which the stream of innate energy becomes silted and bogged. Those characteristics are generalization and smallness. Only one jet of the many currents of life (currents which we trace in the evolutions of the varied species) actually succeeds in retaining throughout the whole time of Life, the essential intensity necessary to travel from unicellular organism to man, and so attain to self-conscious awareness. All the rest fail and we see their failure, their arrest and their extinction marked by their growth in specializations and in accumulation of tissue. We can trace already a number of such ascents, "markings of time," descents and extinctions, clearly illustrated in the fossil records. There we see, time and again, a small, active generalized type begin to lose this anonymous freedom, begin to show specific characteristics. At first these special characteristics seem to aid it in becoming able to resist outside stimulation, to preserve itself successfully in precisely those circumstances in which it finds itself and to stay within that range of awareness to which it is accustomed. Later, these specific characteristics (no longer ruled, shaped and streamlined to a purpose beyond theirs and the whole organism's functions), begin to secede and develop on their own. The great feudal lords are no longer controlled by a sufficiently dynamic despot; they disobediently return to rule in their own territories regardless of the central power, and anarchy results.

Only when the single jet, which rises from unspecialized cell to unspecialized mammal, reaches man, and with that and in that human body attains to self consciousness, at last, in that jet and stem, is the rule broken. Man becomes simultaneously a giant and a speciality, the creature who has at last hypertrophy (the brain) and mass—those two things which all his ancestry, from Life's dawn, had avoided, and so had rendered him possible.

Such a story, if we considered no more than its physical aspect, if we stopped short with the animal evolution and claimed, like doctrinaire Behaviourists, that we knew nothing more of this creature but his bodily biological development, would compel us as naturalists to suspect some further very strange behaviour on the part of the beast. Now that he is about to release and realize the fortune of unspent energy which his vast ancestry accumulated and passed on to him, he is apparently not going, like all the earlier forms of life when they cashed out their inheritance, to develop wings as did the birds, grow fantastic horns as did the ungulates, or even to double the length of his arms, as did the apes, the better to acrobat about in the trees. His specialization is none of these. It is a brain, and so as a first step, he is bound to see whether he can beat the bees at building and the ants at industry. In short this creature's future evolution will be a civilization. The new extension of his limbs will be tools and the shell and frame of his new mind will be houses and towns.

To assert that this is so—that there is no break in evolution between animal and man, between physique and psyche, but that man in his history and culture only, if vastly, accelerates natural history and morphological change, is to be too naturalistic for the idealists and too idealistic for the naturalists. Yet the only solution of the mystery of man lies, it is clear, in such a linkage. To make this fact more easily acceptable we must show that there is in this hypothesis nothing sensational, or if the older word is considered more precise and offensive, miraculous. Physical evolution is paralleled by, is the invariable concomitant of a psychical evolution. Consciousness did not suddenly start out in man like Athene from the forehead of Zeus. There is a mind-body complex present from the beginning of Life and that mind is, at the source, as rudimentary and unspecific as the single cell in which it is manifested. The organism increased in awareness by a series of steps, each indicated by corresponding alteration of physique. The first stage is, we have seen, by the attainment of direction, orientation or polarity; the second, the dominance of the brain over all other nerve centers; the third, the coordination of the body through the development of the hand (and finally single-handedness) as a universal organ continually cooperating with the brain.

The first stage is when the body attains an irrevocable salience through vertebration at fish level. This stage is illustrated to-day by a surviving type, the little lancelet, Amphioxus. This creature seems to represent this stage of development with considerable exactitude for though it has, outwardly, a frill fin running the length of its arrow like body and has, within, both a notochord and a spinal chord, it still has no specific brain and no skull. Ascidians, creatures which still in their larval stage retain such a generalized structure, seem to have been the progressive forms of life in the Cambrian Age (perhaps five hundred million years ago) when life was still confined to the water, and from these Ascidians, or one species of them, it is highly probable a true fish was evolved. It is also important to note that as far down as this stage, of simple directiveness, we can see not only ascent but also retreat. To-day there exist Ascidians which still begin their lives as free moving creatures, but as their individual life continues they lose even the slightest mobility and root themselves, sinking down to a subanimal, vegetative, sessile level. Another path of degeneracy from this very moderate height of simple orientation is shown by the starfish. There is reason to suppose that it is derived from a creature that was elongate and directive and has reverted to indirectiveness.

The next great step is, naturally, among the full fishes and gives rise to a creature of unprecedented mobility and aim—the shark. The long frill fin of the body becomes divided into fore and aft fins, as though an internal energy having precipitated and focused in a head-brain and so, becoming increasingly aware of what it would do, and of the inner drive which it must express, no longer could utter itself in a vague movement along the entire body. It looks as if, through this fingerless elastic glove of flesh, specific sense, a defining craving, thrust out tentacles to hold the water and through it drive the body. Whatever provoked this immense change toward definiteness, it is clear that thereafter the creature that will advance, will cleave its way to its goal instead of wriggling slowly toward it. From this great division of fins will grow fore and aft limbs and in the end, feet and hands.

The following stage is outwardly not so spectacular but it is as significant of radical change. The shark type is succeeded by the Ganoid, fishes which developed out of surface scales a physical frontier made from a mosaic of bony plates. Nor is that all: the internal bones which till then had been cartilaginous began to be true bone. These Ganoids, appearing in the seas, the mud of which when dried forms the old red sandstone Devonian beds (perhaps three hundred million years ago), are important because they give the link between the sharks—the first creatures of limb promise—and the lobe-finned fishes, the fishes which were to show what it was that such a fourfold development promised. The sharks failed. Possessed of this new power of direction and mobility, the energy within them became completely content with this new mastery. They attempted nothing they could not do supremely well. They became the masters and pirates of the sea, the most perfectly adapted of all swimmers, the last word in streamlined finish, the most ravenous of consumers. All their energy went into pursuit and devouring. They remain nothing but incarnate appetite, sweeping the water and consuming all they overtake. Here then, again, we see descent from the line of ascent, as a stage before we caught sight of the ascidian and the starfish types which abandoned even the rudimentary stage of awareness which they had won.

The lobe-finned fishes show how the next step is taken. The fin, which is a hand-bud, must sprout if ever a hand is to be, and grasp is to be attained. The price was big. The shark, making the winnow of his fin ever more efficient to carry him on to his prey, made the same short-sighted mistake of limited efficiency which, long after, Archæopteryx, or whatever was the trial bird, made when it abandoned its potential hand and smoothed down its tentative exploriativeness into the sure, uninquiring sweep of the wing. The lobe-finned fishes had a bad fin, a poor instrument with which to gain immediate satisfaction, safety or ease, but as full of promise as it was immediately clumsy. For, from such a fin would sprout fingers, a gripping hand would result, as from the divided primal fin, divided into four fins, would also spring the limbs in due time. These strange fishes died out, after making the link between the true fishes, creatures whose only possible world was water, and the amphibians, animals who continually attempt to live in two worlds. The sharks still exist, though their giants—some sixty-foot-long fossils have been found—are long extinct. The lobe-finned fish passed into the true amphibian. This is true physical sublimation. By developing this rudimentarily-fingered fin they were able to launch their line toward the shore, and find purchase which would allow immensely accelerated advance toward the goal of understanding and power. By striving into and grounding themselves on a more resistant but more malleable and memorable environment than a liquid one, they can alter their surroundings so that traces remain, outward history begins and they themselves cumulatively are affecting and affected. There begins to be some permanence in their acts; the effect of their movements continues shaping their future behaviour. Their names are no longer "writ in water" and they become reflective, finding themselves continually confronted with the material memoranda left by their past conduct.

A fourth epoch is then entered. The first was when orientation was finally accepted and witnessed by vertebration, when the creature at fish level becomes no longer a diffusive consciousness equally open to the whole circle of stimulation, but pointed in one direction. The second when, following this directional drive, the full fin divides into four specific fins to sweep the creature forward in the direction of its single focused desire. The third when the fin becomes a paw with rudimentary fingers; as the highest fish, no longer content with the yieldingness of water, seeks, however tentatively, to lay hold on the solid. The fourth epoch is when this striving becomes an accomplished fact and, although the new won territory of the land can only be held intermittently, the amphibian, in spite of his periodic retreats into water, can look upon dry earth as the environment in which his true, most present life is lived. The creature which actually made possible the first amphibian is possibly Sauripterus, a fish of the Devonian deposits. The front part of its fin seems to have developed into a crawling hand while the back part remained still a fin. Brain, eyes and limbs all push forward to more difficult but more definite surroundings. Under this stress the whole body seems to have had to make unprecedented adaptations, so that the new life might be possible, and lungs for air-breathing appear at this time. The first amphibians, the first model of a landholder is a minute creature not a span long. So it obeys the basic rules of life, smallness and a generalized structure. This creature is common when the coal measures are being laid down, perhaps two hundred and fifty million years ago.

As this epoch closes and the vast tree-swamps begin to enter the end of their period, life takes another decisive step. The water can now be wholly abandoned and there evolves the first complete land dweller, a creature to whom the water will now be a danger and no longer a refuge. This creature has henceforward all land for his room, instead of being confined to the frontier area of the shore. He moves with increasing agility, and as we have seen, from being a crawler he will become a runner, a leaper and finally (in the winged saurians) a flyer.

Yet all this abundance of inner physical invention must not blind us to the other factor in evolution, the Time element, the fact that there is a tide when advance in physical invention can be made and will lead to a completely new quality of consciousness, but, that if that advance is not sustained and if it is then attempted at too late a time, only failure results. The "composition" which we call the history of life has, like a musical composition, been conducted at a certain tempo, and when one movement is over, those who have failed to play their parts there, can find no place later. After the lobe-finned fishes no fish appears ever to have given rise again to an amphibian, though at least two hundred million years have elapsed since then. After the amphibians of the Carboniferous, no amphibian seems to have given rise to a reptile. Too late the frog Breviceps has completely abandoned the water and the female has to lay her eggs wrapped up in a gel of mucus, but time has passed this "unwise virgin." Its stock will never give rise to higher animals. It remains simply a land frog, not a reptile.

So too, all the Saurians, after their immense achievements, even to winning winghold on the air, fail to persist, and after the close of the Mesozoic and the beginning of the Tertiary, Mammalian age, no lizard will give rise to a mammal. The first great stage of Evolution closes then with the Mesozoic Advance, an advance limited to awareness attained through directional development of the whole physique. Such a line has reached its limit. Henceforward the nervous system, which, we have seen in the Saurians, was still not centralized and therefore incapable of concentration and complete purposiveness, must be the growing point of life. Brain, not physique and limb development, must decide further advance. And, as in order to make the first land advance, which the lobe-finned creature craved, lungs were necessary to breathe the air, so now, to sustain unintermittent consciousness, warm blood from a four-chambered heart must be provided. The lizard can only lapse into coma as soon as the air chills. The mammal will remain uncomfortably, and so questioningly, conscious. A flexible skin, the better to feel and suffer, and hairs which give cover but, unlike scales, signal every touch, these are added features for a heightened feeling. Yet as soon as the proto-mammal, a minute shrew-like creature of some four inches in length, had established itself with its new heightened quality of consciousness, the price of further development once again was found too great for all but one of its descendants. In the first stage of the Mammalian Era, the Eocene, we see, made by mammals, all the physical mistakes which were made by and which destroyed the Saurians in the Mesozoic. In the Baluchitherium we have a mammal which in the Eocene goes through the whole of its evolution, from a small mouse to a giant "rhinocerine" type, standing seventeen feet at the shoulder, and, which having achieved such hypertrophy, becomes extinct. In the whales we have another retreat from life's intense growing point when another Eocene mammal withdraws into the sea and there can hypertrophy, no longer held challenged and shaped, refined and focused by a resistant environment.

In the main line of development unspecialized smallness will be retained and the increased awareness, by the heightened stimulation given by sensitive skin, will be, like an increase of illumination, focused in a higher brain. A mounting volume of awareness will then be available for an ever more discriminating power of understanding. Light and lens will advance together. This creature experiences more sensations than any earlier animal but he is not utterly distracted by these, because he brings to them a new quality of attention. Higher sensitiveness combined with higher mentality leads, instead of to bewilderment, to wonder. In co-ordinating such varied stimuli as sight, touch and sound, the creature becomes with increasing consciousness aware of a world outside his body. The first true mammal, a tree shrew, is still outwardly in many respects like the proto-reptile.

The distinctive feature is in fact not physical. It is that heightened tempo of action, widened response to stimuli, and enlarged range of awareness, in short a new mental quality of being. Such a creature, under such internal urge, begins to evolve so as to satisfy these demands. In the Tarsier we have probably the next distinctive development. Here is still a midget creature—it can sit comfortably in the human palm—with a hand even further removed from the paw than are the hands of any of the apes: with the eyes brought round to the front of the head to give binocular vision and the whole body "mounted" so as to give these delicate hands and exact eyes opportunity to come into co-operative action. From this type must have been developed a manikin forerunner of man—a creature which when it rose to the size of the Gibbons probably added to the tarsier equipment two more powers to explore and to reflect—the power of walking upright, when the hand becomes wholly freed from locomotion to be the eyes' servant, and the power of speech, when the impressions of others can be used to correlate with one's own. To binocular vision, which gives accurate focus and so an extra-tactile sense of volume, is added multimental awareness. So are gained a number of extra physical angles on any problem, which angles, added to the personal impression, lead to an increasing objectivity.

With that stage attained, consciousness rapidly takes on a still further and more specific quality. The humanoid body, now with hand and brain in constant interchange, has in the hand a constant inquirer and in the brain a constant suggester ("Let's try"). We know that at this point, the neopallium, the structure in the brain in which advanced consciousness is centred, stretches out and over covers the olfactory centre. So smell ceases to dominate, as reflection, simultaneously, becomes possible, through precisely that new, higher nervous centre which puts the scent centre out of action. Again we have one of these remarkable convergencies which make it hard not to think that the initiative for such developmental change does not come from the physique but from the psyche. The hand, as a universal joint, will, under this mental stimulation, be ever more constantly picking things up and handling them, prompted by the ever-curious brain, the ever-stimulated body and the ever- "caught" eye.

With this type, physical evolution has reached its limit. The brain rapidly enlarges and reaches its limit, the last organ to take on a specific development. The body simultaneously undergoes gigantism and the creature, which at the threshold of humanhood was two to three feet high, becomes a giant five and even six feet tall. Here, too, ends the advance of consciousness through an undifferentiated subconscious, or rather psyche,[1] expressing itself in tissue changes, manifesting itself through physique.

This was another of the great crises in Life's development and, as always is the case, it was more sudden, more acute and required a more radical solution and invention than any of the earlier ones. When the fish crept from the sea it had to gasp its way into the achievement of lungs while holding on with half-evolved hands. When the reptile cast its scaly skin for good and shuddered in a naked sensitiveness, it had to thrust back the numbing cold through a four-chambered heart pumping out hot blood. When physique could do no more and understanding could be advanced no further by evolving new organs, the humanoid stock had to make a completely new departure. At this point, had not the psychic factor in man discovered a new way to advance him, he would have had to become extinct or obsolete, as did all the "trial" men at this period (the Pliocene or late Miocene—one to four million years ago). Not one of them survived, though there must have been many genera, for we have already found at least five distinct genera—Eoanthropus, Cypanthropos, Pithecanthropus, Sinanthropus and Neanderthalis.[2] Man had either to become extinct or his only other choice would have been to step down and become nothing but a fourth

1. At this stage as consciousness is as yet undivided into self-consciousness and subconsciousness, it is inaccurate to speak of the undifferentiated mind, expressing itself through altering the physique, as a "subconscious." It is not submerged. It is not even emerged. There is as yet not a "land" and a "water" but a primal "liquid." This unself-conscious awareness, expressing itself and its needs directly through bodily morphological change, had therefore best be called the Physique's mental counterpart—the Psyche.
2. All these genera of humanoid stock are extinct though rare individuals seem to have lasted on parallel with man, such as Eoanthropus II, dated in early Paleolithic levels, and Neanderthalis which lasted on in large numbers till the middle of the Paleolithic; while Cypanthropos skull was found associated, at Broken Hill, Rhodesia, with nothing but extant animal species.

party to that dreary trinity of degenerates, the Gorilla, the Orang and the Chimpanzee. Such human decadents already exist in those forest haunting families—they can no longer be called tribes—which have lost almost all their crafts and culture, now occasionally come upon in the interior of the Malay peninsula, Celebes and perhaps in Borneo.

The new way, the radical psychical mutation was to shift from change of tissue—Physique—to the incomparably accelerated change in consciousness which is caused by all change being able to be expressed henceforward in Technique. We have seen, it took some thousand, five hundred million years (during half of which living forms were still without even orientation) for Life to rise from unicellularity to pre-technical man, the creature without tools or crafts. This is one self-consistent rhythm, one single method; mind working itself out and finding itself and its setting through physique. At this level any wish to extend apprehension can only be gratified if and when an organ can be grown to supply that apprehension. Any desire, sufficiently long sustained, will lead to such an organ as may express it. But any desire, short of wanting to know all, will prove fatal to further advance, for the local wish will block the larger wish and the good prove fatal to the best. Yet all these shorter wishes will be more easily and more quickly satisfied than the larger, undefined, hardly apprehended desire, only comparable with the migrating urge among birds. Only those enduring to the end are saved: only those who are never satisfied with anything which would sate their present nature will find the inexpressible experience that lies beyond.

The next stage is immensely shorter in time and prodigiously accelerated in tempo. A thousand, five hundred million years is shortened to perhaps three million years—at the longest computation—an acceleration of five hundred times. And even that is to extend vastly the Technique Epoch, by expanding to the utmost limit, the Eolithic Age. Most students would not extend it beyond one million. (But we must remember that a pierced and shaped tool of extinct elephant bone was found with Eoanthropus I. This shows much power of purposive concentration: Pekin Man (Sinanthropus) was said at first to be without tools or crafts and now is found to have had flint knives, fire, and animal skull cups and horn tools.) However the dating of this almost accidental technique of the "dawn stone" culture may be settled, it is clear that that culture which succeeds it, the Chellean, the first age of fine chipped stone tools, is dated now only five hundred thousand years ago—a mere moment and beat in the tempo of physical evolution. Concentration to carry out a mental image, a conception; the power to hold a picture in the mind so as to shape a purposed object, that new type of Psyche, of awareness, of evolutionary power is only half a million years old, and so the acceleration in this second phase as compared with the first would be three thousand times faster.

To-day we are at the third great stage and must face another more profound change, as evolution strives to advance consciousness still further and to create still another quality of awareness. We are self-conscious. The primal consciousness, with the specific rise of man out of animal, became divided into an aspect devoted to the maintenance of the body—Physique—and an aspect wholly concerned with the mastery of the environment—Technique. This leading off of creative interest to activities outside the body meant that the physique ceased to evolve, merely maintaining itself, merely "marking time." Evolution, however, did not stop there, with man having become *homo faber*, man the technician. He had to attempt to become *homo sapiens*, man the comprehender. A further division and extension of consciousness takes place when psycho-technique, that awareness of purpose which leads to the creation and development of crafts and material instruments and finally to deliberate mechanical invention, is succeeded by what may be called the psycho-psychic, the awareness of the self, of the nature of its purposes and of the meaning of Desire.

It must be repeated, it is very hard for people to believe that this is the choice before them. They have for so many years fluctuated between the vague hope that after all "An increasing purpose runs: And the thoughts of men are widened, With the circling of the Suns" and a general irresponsible assumption that the whole thing has no purpose but that, as we only live once and as evolution can only be inconceivably slow and even our societies and states take a full generation for any radical change, even the decline will last our time. As in the Falling Nightmare, we shall never live to complete the fall and hit the floor. The vague optimism that "evolution is going on" is denied by all present evidence. The unconscious development of man is ended. It closed at least a million years ago, perhaps three million. The alternative optimism that "progress" is going on and that our civilization cannot crash because its technical advance is spontaneous, inexhaustible and beneficial, is also contradicted by experience. That advance of evolution, which is here called psycho-technical evolution, has also ended. It ended in this generation. In our time the attenuated association between means and ends snapped. That association is an essential connection for any further progress. To-day we have the technician, because his means are divorced from ends, destroying civilization (and with it his technicality: for physical science is destroyed by the scientific war it makes possible) at the command of men below the technician in any mental calibre. The final despairing selfish hope of the cynic that, though the ship is sinking, it sinks so slowly that there is plenty of time to drink oneself to death before the water pours in, is denied. For evolution has now passed up with a sudden change into the third acceleration so that alteration is now so fast that it is apparent to us. It becomes apparent to us because evolutionary change is now as fast as the age-change of our individual bodies, for that evolutionary change is no longer working through change in our physique or our technique, our features or machines, but through ourselves, through change directly in consciousness itself.

Even if we had no actual contemporary political observation that this were so, we should be able to deduce it. From the common principle shown by a number of intellectual developments we should be able to know, without any news of the present world, that man must be at the greatest crisis of his evolution; as a mathematician can calculate, without looking at the sky, the conjunctions of planets, or an oncoming eclipse.

Briefly these are the symptoms of oncoming crisis only to be solved by a psychic evolutionary mutation:—In the world of action there are, (1) the accumulation of means at the cost of losing adequate ends, (2) the acute division between reality and ideals, (3) the loss of sanction in civilization. These are a recapitulation in terms of society of the degenerative physiological symptoms which mark the end of an animal species; the fatal accumulation of tissue and uncoordinated specialization of parts.

In the world of thought the same condition has been reached. In physics the new quantum theory ends mechanism and returns the problem of reality to psychology. In physiology the new concept of Apprehension—the evidence of our insensibility to, but affectability by, indefinite octaves of radiation, ends all assurance in the finality of Common Sense. Heraclitus was right, though Thales won approval; the senses are bad witnesses. In metaphysics the new concept that Time is only an aspect of Space Time, and non-existent without its invariable associate, shows that our cosmology now requires a Copernican revolution in our idea of Time to complete the earlier and incomplete Copernican revolution in the idea of Space. Time, we now realize, is still and we move through it and if we would understand ourselves and how we affect things (which we thought affected us), we must realize this fact and act upon it, as clearly as we realize that it is our motion which makes the fixed stars appear to move.

It may be said, all these ideas have been known already. There is nothing mutational about them. Their emergence does not show that we are suddenly on the brink of racial change. True, they have been known to the few and lived out by the very few, rare psychic pioneers, creatures as much ahead of their race as the proto-mammals of the Cretaceous, were by millions of years, forerunning the Mammalian Age. But to-day all of us, as a civilization, a society, a humanity, we, the race of man, enter the Tertiary Age of our specific evolution—an age as different from our Secondary Age, our age of Technique, as was that age a radical departure from our Primary Age, the age of evolution through Physique. And as our Secondary Age of Technique showed an immense acceleration over the rate of change possible in our Primary Age of Physique, so this Tertiary Age will develop at a proportionately accelerated speed beyond that of the Secondary.

There seems then no escaping the conclusion that nothing less than what can only be called a mutational change in the quality of consciousness itself will answer our present need, solve the problem with which we are confronted, and prevent our lapsing into decadence—at our rapid tempo of psychological advance shown and showing itself as individual and mass neurosis. "Inspired common sense," action, however intelligent and devoted, *on the present factual findings* and attempting to bring them all together and from the sum to find the basis of a rational behaviour, that is no longer enough. That must lead to complete breakdown, just as much as the most selfish action; indeed perhaps more speedily now, for a number of people all acting on animal self-interest will slowly destroy each other, while whole masses acting devotedly on wrong and self-contradictory principles can make destruction almost instantaneous. This is because, not merely morality has broken down, but common sense has broken down. That consensus of the senses which gave man, up till now, his sense of knowing what he was experiencing, what he was up against—however he might behave—that basic assurance that his experience, the data of his consciousness, did make sense, no longer as a matter of fact, exists, for the findings now contradict each other. We have to realize that this difficulty is so deep because this problem goes below morality. Though it is already giving rise to the ugliest moral consequences, such as chemicalized slaughter of cities, war to the utmost and without any psychical restraints, it goes deeper than conduct and the deduced behaviour which we call ethics. It rises from our own consciousness now being unable any longer to make a single consistent sense of its undoubted experiences. Science, the pure wish to understand, is just as bewildered as is politics, the wish to maintain order and liberty and to find an enforcement for them which will not utterly destroy what it is intended to sustain. It is not merely that the conclusions we draw are wrong and lead to evil but that the facts which we perceive do not themselves make sense, cannot be made to self consist and yield a comprehensive meaning, however inhuman.

The nineteenth century found itself faced with the serious problem of carrying on human society in the new belief that the Universe was a strictly necessitarian machine and Life was a struggle which must go to the most violent and cunning. The two solutions that resulted were (1) that man should accumulate such resources, by understanding nature; that within an artificial enclave, so made possible (i.e. civilization), he could suspend Natural Law and live according to his finer feelings; and (2) that man should obey Nature and make himself, by the utmost militarism, the master animal over all his fellows. Before, however, either of these conclusions could be given more than a preliminary try-out, there disappeared the common basic belief on which both plans were founded.

The twentieth century faces therefore the completely different, far graver and more difficult problem of carrying on human society in the belief that Nature itself, the physical world (or our detached apprehension of it) is basically inconsistent, that no single principle explains experience, and that the varied principles by which experiences are "explained" (be it the behaviour of matter, as the quantum theory leaves it unresolved, or the behaviour of the brain, or the behaviour of man—human conduct) contradict each other. This paralyses all action, however ruthless, and that is why to-day we see the mechanized militarist with his out-of-date realism being superseded by the fanatical hypnotic "sleepwalker." The man who can be most madly visionary wins all along the line. So Hitler wins on his home front against his practical generals, and on the foreign front against the practical Chamberlain and Daladier. In a world of fantasy the fanaticist will outstrip any who try to be rational.

The truth is that we are, in this generation, passing over the threshold from one Time Sense into another. Indeed, life's development, the evolution of awareness, may best be rendered into three attitudes toward Time. Then we see successively, the Time-unaware animal, the Time-haunted man and the Time-understanding mind. To-day we are suffering from mental and moral vertigo, because, while still mainly aware of stage two, the third stage is dawning on us.

Consciousness must then evolve. No lesser, more partial development can save us from swift collapse. Otherwise we have reached the point when an acrobat finds that the horses he straddles are bolting apart and he is being flung to his death. We must have direct and definite enlargement of our experience of reality, in which enlarged experience our present partial and conflicting experiences may either be seen as misapprehensions or seen to take their places as reconcilable with each other in the reconciling synthesis. We must by this radically new evolutionary mutation, win to a clear, "objective" apprehension of extended Reality, a reality of which the animal and the animal senses are wholly unaware.

That being the next evolutionary step, we at this moment being come to it and being driven to extinction if we cannot rise to it, there remains to ask the following questions. Before we can decide, however desirable, however necessary a change would be, before asking, "Is this possible for us, or impossible?" we must first ask, "Is there reason to suppose that there is still such a sum of 'raw' unused evolutionary energy—sensation-capacity energy, above that required to keep us fully conscious—out of which this new faculty of apprehension, the new quality of consciousness might emerge?" That is the question examined in the two succeeding chapters. Once that is settled, that there is evidence of such a store still remaining unused within us, we can consider the evidence for what "growing-points" have already appeared; growing points from which the new departure might arise.

If man's evolution is to become psychical, from having been physical, ought we not to be able to detect a middle linking phase, when his evolution became psycho-physical, when, between the evolution of organs and the evolution of consciousness, the evolution of senses took place, when between the animal's outlook and full consciousness, direct awareness, the animal senses were enlarged to their limit so to serve a more-than-animal desire to apprehend? If that stage can be detected, as Chapter IV would indicate, then it greatly strengthens the belief that beyond that stage lies the further condition of full consciousness.

CHAPTER II

THE UNUSED ENERGY REMAINING FOR FURTHER EVOLUTION

(A)—The Function and Future of Pain

IN THE last chapter it was asked: Is there evidence of a sum of raw energy remaining in us? The presence of such a force, if it exists, would be detected by two symptoms. If such an immense power remains unused in man and if, as we have seen, it cannot find its adequate outlet and sufficient expression, unless we understand the purpose for which it exists, the development toward which it points, and so co-operate with it, then as it is balked it will give grave trouble. The symptoms of that frustration will be, in the individual, physical distress and physical excess and in the human society ruthless competition and war.

Of course, if such an idea as the continuance of evolution and that continuance being psychical, seems impossible, then the evidence, which suggests that our nature contains the energy for such a further advance and that, thwarted of its true purpose, such energy must and does give trouble, will be read differently. Social conflict, war and even individual neuroses will be explained, for instance, as due to economic conflict or to social regulations inhibiting, and so rendering morbid and malignant, urges which, given expression, would be harmless.

Severe physical pain is however more difficult to bring under strict Marxian or Freudian explanations. The acute agony which is experienced when striped muscle is crushed or damaged and when any pain nerve is pressed upon is not only the intensest feeling of which any ordinary man is capable but it is also an experience which none of the present popular interpretations of sense data can satisfactorily explain. In fact, in spite of the general acceptance which has been won by the Marxian or Freudian interpretations of Life, if we are to have even a very moderately successful comprehension of the intense, frequent and baffling sensation of pain, we still fall back on even more rudimentary attempts to understand this, our most vivid and least comprehensible experience.

That we should still be incapable of providing a more modern explanation is certainly strange. We have yet to believe in the evolution of consciousness and so we cannot but assume that in pain man has always been confronted, as keenly and inescapably as ourselves, by this the most acute of all human problems. It is therefore clear that wonderfully little progress has been made toward explaining it. Though so spurred by his body, along this line man's thought seems to have been singularly unable to advance. During the ages in which man's outlook was anthropomorphic he took for granted that suffering was divine revenge. Then, as his view of deity became nobler, pains were interpreted as penalties exacted for infractions of the moral law. When such an infraction could not be established, pains were held to be trials, tests to establish whether man could rise superior to his own distress and not deny the righteousness of God, although he personally suffered: "Yea, though he slay me, yet will I trust in Him."

These apologetical efforts, it is clear, were never very convincing. As soon then, as anthropomorphism was superseded by mechanomorphism and the ultimate explanation of things was postulated to be a magnified, non-natural machine, instead of a magnified, non-natural man, the problem of pain was interpreted in terms appropriate to the new cosmology. Even as early as Bishop Berkeley we find this Christian and Idealist philosopher declaring that there is no problem of evil and pain. For as the whole outer world is simply an immense machine, running faultlessly by obvious natural laws, God teaches us to understand it by permitting us every now and then to pinch our fingers in the wheels. Pain is nothing more than the quickest way of educating man in objective knowledge, and, this kindly Bishop maintains, compared with this treasure of scientific understanding, occasional agonies may be dismissed as insignificant. Pain is a cheap method of buying essential information about the nature of things.

The shadowy God of Berkeley and the deists shrunk like an evaporating cloud behind the rising mountain of mechanomorphism. There was no longer any need of the hypothesis of the mind underlying matter and of the lawgiver informing natural law. Pain began to be viewed as automatic as the rest of the physical working of things. And to-day, although mechanomorphism is now itself a fading picture of reality, nearly everyone, educated or uneducated, believes that science has once and for all answered the question: What is pain?

It is simply the body's burglar alarm. A nerve, when it is tripped over by an intruder sets ringing a signal in our brain. So we are made to attend, eject the trespasser and put the damage right. Then the alarm signal stops. That undoubtedly is partly true, though a half-truth which by giving us semi-understanding has given us only a partial, arresting power. We have been able to prove that pain is a call to restore an infringed freedom, by stopping in this way many sorts of agony. From the earliest and crudest tooth-pulling to the latest methods of checking sciatica by injecting alcohol into the nerve channel, we have had dramatic demonstration that the nerve carries the pain impulse, that pain arises from some bodily damage, and that if we inhibit the nerve we inhibit the pain. Indeed, so remarkable has been the advance in pain stopping during this century that many doctors and physiologists now say that we may get rid of pain altogether.

Not only will anesthetics pass us through all acute crises of agony and the analgesics take the edge off every ache, but such chronic tortures as *tic douloureux* will yield, as this particularly frantic neuralgia has seemed to yield to a drastically corrected diet—in this case immense doses of Vitamin B.[1] When we have relieved every pressed nerve, when we have fed to every fibre the full supply of nourishment, lacking which it has become liable to such frantic spasms, then every-

1. The California Institute of Technology is supporting the Clinic at which some very promising results have been obtained. After keeping patients for a month or more on a dose of Vitamin B, ten times in excess of that needed by ordinary healthy persons; of the twelve who were experimented on, eight patients declare themselves free of the pain and the remaining four say that the pain is now quite bearable.

one may have a body which, as long as it is handled with high intelligence, would never pain.

Yet it is precisely this research which, in getting rid of many sorts of pain, has revealed to us that pain itself is much more mysterious than we had assumed. The simple mechanical explanation of it, research shows, is not enough to cover the actual facts. For example, the study of pain has shown that many nerves frequently "pain of themselves."[2] Certain violent neuralgias often come on as punctually as a sunset gun, and no physical cause can be discovered for them. An even more serious fact is that the more we learn about new methods of stopping pain, the more pains there seem to be. Dr. Carrel, the famous Rockefeller researcher, a physiologist whose discoveries about the nature of living tissue are among the most remarkable of our time, has given us a grave warning in his book, *Man the Unknown.* He points out that much of our so-called progress toward comfort, security and painlessness is not really an advance, but a regression. We are only exchanging a natural, automatic defence for an artificial and uncertain one. For the body, when it lives under natural conditions, has enormous capacities for enduring hardship without damage.

Take the most obvious of all discomforts, the dread of cold and the damage it can do. To guard against this, we wear layers of clothing and live in heated houses. All that this does for us is to make the body's "vaso-motor," heat-controlling system cease to work. We have then exchanged for a natural automatic defence against the pain and misery of cold, a clumsy, "economic," inaccurate defence. We are therefore only the worse off for our progress. How much we have actually lost our power of protecting ourselves by so depending on clothes, et cetera, is shown by Dr. Rollier's sanatorium children at Leysin, in Switzerland. These children have all been tubercular, but by re-training their bodies' natural capacity to stand cold, they are made so healthy that not only is the disease itself cured but they go out regularly skiing and need wear nothing but their skis! Nor is this the limit to which natural resistance to the pain of cold can be developed. Reliable anthropologists now confirm the report that in the Himalayas there are men who

2. See Dr. Balme's *Pain and Its Prevention.*

train themselves to be able so to hold off the intense cold that they can sit in the snow for hours, wearing nothing but a cotton robe.[3] In short, as soon as we look into this problem of pain we find that we are up against the problem presented by the Princess in Hans Andersen's fairy tale. The softer they made her bed the more she felt, until one crumpled rose leaf under twenty feather mattresses seemed to her like lying on a sharp pebble.

Indeed, it seems that we may have put the cart before the horse in this question. We were sure, with our materialist cart, that pain always was a consequence and could not be a cause. Tissue was damaged and then pain resulted. Leaving aside for a moment the fact, which will be noted later, that many of the highest centres of the body can be damaged, life can be in mortal danger and no pain felt, it is increasingly clear from increasingly accurate diagnosis that many pains do cause physical derangement. Examination at what is among the world's five greatest surgical clinics has revealed that ninety per cent of patients, who feel in such gastric pain that they are ready to undergo the long and thorough examination and pay the large fee such an institute charges, have, at least as yet, "nothing wrong with them," no organic lesion. However, now they are not dismissed with this careless, inept comfort. Pain is always real, even physiologists now realize, though its cause may not be physiological. If left so, the patient's pain will continue to work until it causes physiological change. Its cause, while still psychological, must be diagnosed. The patient is told to watch his way of living, asked whether he is worrying, whether he is anxious and thwarted. Such advice may not, indeed seldom can, arrest pain and make it cease to gnaw till it destroys tissue. The patient, as one of these diagnosticians remarked, generally needs no operation, no medicinal treatment. But he does need something very badly and, if he fails to get it in time, he will be back in the surgeon's hands, this time with something that a surgeon can see. He needs a different life, a life of purpose and of meaning to raise him out of his growing and gnawing sense of futility, frustra-

3. See Dr. W. Y. Evans Wentz, *Tibetan Yoga and Secret Doctrines,* Oxford University Press.

tion and inevitable failure. Pain, then, is going to prove a larger thing than we thought and a deeper thing to eradicate.

We have to ask, then: Is the explanation, "Pain is a Warning," a true answer to our question? True, it may be a warning that we ought to do something, but our mistake seems to be that when warned, we have done precisely the opposite to what our deeper nature intended. We have decided that Pain is a warning, and a warning which has only one meaning. We are to find and repair some bodily damage, to make ourselves comfortable, to restore the *status quo.*

Now the first thing to be noticed in that explanation is that it does not cover even half the facts already well known. We have in our own body clear proofs that pain is of little use to us as a warning against physical damage. The first of these proofs is external, the second internal. All over the skin there are distributed beside the endings of the pain nerves, those of the pressure-recording nerves and also those which indicate heat and cold. We have, then, a much more delicate and informative recording system than pain itself gives; for those other sensory systems respond to a far slighter warning than the pain nerves require before they will register. The second proof that pain is inadequate as a warning is even more conclusive. Many of the gravest pains are internal. Yet who has ever been warned from having appendicitis through an attack of colic: does nephritis help the body to avoid forming stones: has anyone been able to cure himself of earache by drawing back from the ache, as we draw back our finger from a flame? It is clear then, that where on the surface, pain may sometimes warn us, we have far better alarms which are not painful; while internal pain, which is often the most violent, can very seldom serve any cautionary purpose.

So much for evidence from the human body. If we search outside ourselves, the facts are even less in keeping with the current assumption. It is taken for granted that as animals in a natural state do not suffer from nerves, any pain they experience must be a clear warning for them to avoid physical damage which would otherwise prove fatal. This argument, although it falls in with a popular conception of evolution, is difficult to establish in the face of evidence well known to most naturalists. Animals are frequently found to be suffering from many diseases—pains from which the lash of their suffering cannot drive them. They are frequently reported as showing severe distress from stone, from abscesses in the teeth and in bone cavities, from osteomyelitis and from arthritis and many other painful complaints.[4]

Often, however, vivisectional experiments on animals have been cited as demonstrating that the *d'être* of pain is to warn sensory creatures. The most commonly quoted example of this procedure is the severing of the pain nerves in the forefeet of rats, after which the animals are reported to gnaw off their numbed limbs. Lacking pain, it is argued, all sub-rational creatures would not be able to resist the temptation to make meals of themselves. This reasoning appears to be anything but conclusive. Indeed, it would seem much more likely that the rats, who have an advanced sensory system, feeling their forelimbs to be strange would nibble at them in a dazed attempt to find out what was this strange affliction. The child who scratches at a pimple is not seeking for nourishment but investigating an anomaly.

The most unsatisfactory element in this, the naturalistic—or what it is more accurate to call the "arrested evolution" explanation of pain—lies, however, in the fact that if pain really served as a warning, then animals would suffer more than men. Having less intelligence to tell them of danger, they should be given a sharper spur to drive them out of its way. Yet the contrary is the fact. It is we, not the animals, who have the greater capacity for pain. It is clear, if the popular opinion were correct, we should have the less. The more intelligent we have become and the more we have learned to understand, the more we should be able safely to dispense with pain. But, as a matter of fact, we have only become more sensitive to suffering.

So strange a fact as this would undoubtedly have incited investigation had not the fundamental assumptions of the time made us disregard it. We are not without clues pointing to a solution. There are already two lines of evidence which are well known and throw con-

4. That this is no modern morbidity, see Dr. Swinton's study on evidence of disease in fossil animals.

siderable light on this apparent anomaly. The first comes from modern anthropology. There used to be a comforting assumption that savages were far less sensitive to pain than were the civilized. The first time, however, that accurate psychological methods were used to measure and compare the savages' capacity for suffering, it was found that his sensitiveness actually surpassed that of the educated European.[5]

This evidence, which seemed inconsistent with common sense, fell into place when it was linked with the finding of other psychophysical research, which had also produced unexpected results. This was the specialized study of the sense organs of educated Europeans, particularly of hearing and sight. These results show that both ear and eye, because they are highly advanced sense organs, have a peculiarly significant way of reacting to injury. Though it is true that both can pain intensely, pain is not their sole reaction. In a large number of their specific diseases the reaction is not one of pain. The ear transmits news of its injury by making warning sounds, and the eye by impulses which the brain perceives as warning flashes.[6]

If we place this fact beside the one mentioned above—that savages have a greater capacity for pain than the civilized—it is possible to see a new association. All mankind has a far greater degree of vital energy than any animal. The more we study savages the more we realize that they are nowhere near the animal level.[7] The difference between them and us lies in the different degree of use which they and we make of this high degree of mental energy. Briefly, they have more of it pent up unused within them. We have begun to draw it off and to employ it. There have been plenty of sociological findings to confirm this. It has been frequently found that savages and even peasants, brought into the intense rush of modern urban life, quickly break down and die. Conversely, only weedy wiriness derived by the drastic selection of city life (and not the agricultural labourer, who was rapidly "shell-shocked") could stand the acute nervous strain of

5. The Cambridge Anthropological Expedition to the Torres Straits.
6. See C. S. Myers, M.D., *Psychology,* Cambridge University Press.
7. cf. Spencer and Gillen's *Arunta: a Study of a Stone Age People,* and other studies of the Australian Aborigines, agreed to be the least advanced of all human types.

modern war without excessive neurosis.[8]

If then the accepted explanation of pain fails signally to answer many of the obvious questions raised by this urgent experience, we are compelled to put forward another hypothesis, even should that prove to be not only revolutionary, but what we have seen to-day arouses even more suspicion, hopeful and constructive. We have to suggest that the evidence indicates that pain is serving a purpose far more important than was imagined. In a phrase, we may say that all pain is a measure, if a crude one, of the degree of vital creative energy in any creature, animal or man. If that creature is an animal, it will suffer little, because however injured, it has only a low vital charge to rend it. For, as we have seen, all animals are living fossils. The life in them is gradually sinking, like a failing lamp. Their evolution is over. Man alone can still evolve. Therefore, if the creature in pain happens to be a man, because of the high degree of vital energy still remaining in his species to provide for its still unfinished evolution, he must suffer intensely when injured, unless he has found appropriate channels through which to let that pent energy express itself.

So the animal has none of this pent energy and therefore feels pain least. The savage feels most because he still has a store of such energy but has little means for releasing it: he himself is not a living fossil, but his society in which he might express his creative energy is fossilized, or, to use Moret's phrase, anchylosed.[9] We feel more than the animals but less than the savage because we have already begun to tap some of that store when we concentrate on creative work.

It need hardly be said that such a hypothesis is as complete a reversal of the common-sense view of pain, as it is of the scientific. It therefore needs far more careful confirmation than would a theory which did not run counter to current opinion. For it is clear that if this new hypothesis is true, it holds out a new hope, not merely for our painful selves, but for the future of the human race. Traditional and current opinion assumes that man's evolution is over. Any sen-

---

8. See MacCurdy's *War Neuroses.*
9. Moret and Davy, *From Tribe to Empire*

sation he experiences, whether painful or pleasant, can only be for conservation and comfort, for restoring a disturbed stability, for keeping him where he is. Acute sensation cannot be intended to spur him to creativeness and to urge him to a new level of being. It must then be shown that there is unmistakable proof that we are developing actual capacities and faculties which so draw out and employ our vital energy that when this is done we are rendered painless. Strange as it may seem, enough evidence has already accumulated to make this highly probable. We have been kept from realizing this because the facts are drawn from many unrelated lines of research. Meanwhile, for want of a better explanation, the old mistaken assumption as to pain's meaning has hung on in our minds, although proofs on which it rested had broken down.

The new principle is simple. It may be stated as a proposition: The more mentally active anyone is, the less is he capable of pain. The proposition so stated, the proofs rapidly fall into place. To start with the negative proof: Everyone has found out to his cost that it is when trying to get to sleep, when relaxing, that pains which were hardly noticeable when the mind was actively interested, now become almost unbearable. It has often puzzled doctors as well as patients that toothache, neuralgia, sciatica, eczema, et cetera, become most troublesome when the body is resting quietly, and the sufferer's only wish is to fall asleep. Yet, if pain is due to our failure to draw off our pent energy by active interest, then it is quite natural that when that interest sinks, latent pain becomes active. It is when we are least concentrating that we most easily fall a prey to physical discomfort and misery. If that is so, the converse should likewise be true: When we concentrate most, we should feel least. This is the positive proof, and it has been as fully established as the negative. The half-awake suffer most; the most intensely attentive are least aware of pain. It is a commonplace of observation that when concentrated on watching anything which excites our interest, looking down a telescope, at a fight or at an accident, we may be burnt or even gravely injured and yet not notice it.

However, the complete proof that pain cannot exist if the mind fully concentrates its interest on something else, is given by hypnotism. The surgical work of Elliotson, Esdaile and Braid established a century ago that once hypnosis is induced the patients become so concentrated that major operations such as amputation of the leg can be performed, and they pay as little notice as though they were under a deep anæsthetic. Yet it is now recognized that hypnotism is no more and no less than the power of being able to hold the entire attention of the mind persistently on one thing, so that there is not a moment's thought left for anything else. Such an intense, undistracted, indeed, unreflective attention evidently reaches a pitch at which there is not only no memory of the last pang nor apprehension of the next, but the pain-impulse itself can no longer register in a consciousness which has become completely instantaneous. There is no longer that threshold and span of the "specious present" (usually with educated European stock some one to one and a half seconds) without which and below which the pain impression cannot be registered. As when the magnification of a microscope becomes sufficiently high, the light wave itself becomes divided up and in consequence only darkness is seen, so sensation itself can no longer be experienced in so high a field-of-attention. The mind has reached a tempo at which the slow beats and throbs of the body are no longer recorded.

That hypnotism is precisely this, is proved by the fact that people can train themselves into states of concentration in which their attention is exclusively confined to the one object which they have resolved to notice. In this way not only martyrs and mystics in the past, but men to-day can and do train themselves to be able to undergo sufferings which the ordinary mind is convinced must produce agonizing pain. It has been shown that the Sufi dervishes, who pierce their cheeks, breasts and thighs with daggers, and place red-hot irons on their tongues, feel only a pleasurable sensation, while if the state is deepened, all physical sensation disappears. All this, it is very important to realize, has nothing to do with stoical strength of will to endure pain. Indeed it is the reverse. No effort of will is made. It is simply that there is no pain, because the current of vital energy has been drawn off to yield the high intensity of attention.

We have not, therefore, to discover new drugs in order to free ourselves of pain. New analgesics can only do one or other of two things: either they dam up the vital energy until sooner or later it bursts out, all the more devastatingly, or they actually destroy that vital energy and leave us without the will to live. For our capacity for pain is no more and no less than the crudest indication of our capacity to produce new powers, to evolve further. Here, then, we are confronted with a stimulating paradox: to aim at getting rid of pain is the worst way of making this attempt, and alternatively, the best way to get rid of pain is to know how to disregard it. This knowledge, however, is in no wise negative. The aim is not to get rid of pain but to achieve a new quality and intensity of awareness, one of the by-products of which is the disappearance of pain. The sufferer does not wish to be without pain and strive to wish his pain away. He begins to understand how to be fully alive, how to be conscious in a way which uses up the energy which was leaking and discharging in pain.

The new outlook perceives that if your objective is simply to become comfortable, you only increase the possibility of pain. The problem it sets us is solved when we use up the raw, pent energy in new capacities. Anguish is simply the shock we feel when that store of unused, raw energy is suddenly tapped. If you grasp an iron handle when the thermometer is well under zero, you feel as though it had burnt you. And this is true: the heat suddenly drawn out of the body has burnt the skin, just as heat suddenly put into the body bums it. If a man is put on a glass sheet and then charged electrically, however high the charge, he feels nothing until someone not on the sheet touches him. Then, at that point, the charge leaves the body in a flash. If it is high, he is killed. Pain, we now know, is like every other sensation, manifested by an electric current. If we can prevent this charge being "detonated" and release it in advanced, orderly activity, we shall become painless. What is even more important, when that is done we shall be more alive. For there are two ways of being painless: one is by being paralyzed or numb; but this other new way is being so actively intense that all the pent energy which could have been wasted in suffering, expresses itself in a radically new form of living.

The problem of pain becomes, then, one that is no longer baffling. Though it remains to be solved, we can see in what direction the solution may be found. Pain is a violent intermittent warning that we should spend and not hoard the most valuable gift any being can possess, the energy required for further growth in awareness, for creative, evolutionary development. We are not at the end of our achievement, but what we take to be a baffling, agonizing thwarting of our desire to live in healthy ease, is the goad to urge us to enter on an infinitely richer existence. Pain, instead of being the darkest of problems, becomes an immensely hopeful symptom. The most dreaded thing about human suffering is expressed in the word itself. Here is an arresting, engrossing experience, about which the victim can do nothing. He sinks to be no more than a patient, one who can only react to the assault at best by patience: he is reduced to being completely passive—all reply, all action is denied.

Having, however, made this discovery about pain, we shall not master this grave symptom unless we act on our knowledge. Indeed, we may now say of suffering—with more truth than could be said of any underlying disease—prevention is better than cure, indeed, in this case prevention may be said to be the sole cure.

There remains, then, for us to find methods of drawing off the primal charge and energy. We have to find and follow this further path of evolutionary development, in which our creative capacity may flow. We may say of pain as Shakespeare said of thwarted, passion, that though when unbalked it is like a gentle stream, "the more thou dam'st it up, the more it burns."

We shall not, however, be able to advance our explanation in this unfamiliar direction, we shall not be able to prove that pain can be eliminated by the further advance of our evolution, unless the fact that there is a further evolutionary goal can be established by proofs other than those drawn from the nature of pain itself. For two reasons most people would reject the case presented so far. They would instantaneously doubt both the evidence and the proof, the goal as much as the means. The very fact that this hypothesis presents us with a doubly hopeful idea (that the worst fact of our present lives may be removed, and that our future holds unsuspected possibilities) is sufficient to make most people reject it even were the evidence such as would silence any rational objection.

The evidence of our capacity for further evolution (in distinction from our need so to evolve) has so far in this book been drawn only from an examination of pain's nature. This presents a peculiar difficulty. Pain is frequently so terrific that when we are offered an explanation of it, an explanation which would give it meaning, we are filled with incredulity. That pain should be eliminated by further development of consciousness seems the most obvious self-delusion and wish-fulfilment fantasy. Any explanation of a thing so terrible would be suspect. When that explanation proves to be one which not only gives pain meaning, and gives us the initiative toward it, but also gives meaning to life as a whole, then our doubts are redoubled. The proposition that evolution may still continue in us and, further, that that evolution may now be through an enlargement of consciousness, is itself unacceptable to all of us whose minds are still restricted to the material explanation of reality. Even if we were not made suspicious by pain being brought into the question, the simplicity and apparent finality of mechanism has closed our minds to further speculation. We feel that it is axiomatic that if evolution continues, it can only be physical evolution. We may perhaps become more effective animals. Our senses might become keener, our brains might become more cunning. That the focus of consciousness could enlarge, that mind could cease to be merely ancillary to body and become the aim of evolution, that is too large a break with the current of contemporary thought. Objective meaning, the postulate that man cannot merely understand the process in which he finds himself playing a leading part, but that the plot may entail the further development of those powers which he feels most peculiarly human, such a notion that consciousness is the meaning of life, is one which is still dismissed as hopelessly anthropomorphic.

Nor do we reject a more hopeful outlook only on the ground of its intellectual inconvenience, because it would upset our simple, plain and limited ways of thought. As we must repeat in order to discount prejudice; quite as much because our feelings have become used to the Larger Meaninglessness as that our minds are at home in a mechanical world, are we unwilling to accept any notions which would ask us to change our outlook. Emotionally we are now clinging to that sense of separation between ourselves and our values, and the universe and its realities, a sense which was acutely painful to our grandparents, but with which we have become familiar. We are far from ready to welcome the return of hopes which we have long learnt to do without. Liberty appears very different to a captive of thirty years standing, than to one beginning his imprisonment. The light which to a man in the dark is a godsend is a bewildering embarrassment to a mole. Doctors have frequently been surprised at the small relief and even smaller gratitude that is often expressed by those who after many years have had their sight restored, or who have been cured of a slow mortal disease. Life is always adapting hastily to what it has, often even more hastily, taken to be the inevitable. To those long numbed by a secular despair (especially when through these years of idealism's winter, private property has kept them shut up comfortably by their own well-furnished hearths) the return of hope is as painful as the recovery of circulation in a frozen limb. Probably hibernating animals experience little pleasure when the weak sun wakes them from warm sleep to hunger and restless prowling in a half-thawed world.

It is not to be expected, therefore, that proof of our capacity for further evolution could be accepted through evidence drawn solely from the nature of pain. Even if by advances in hypnosis anyone's sufferings could be stopped; even if everyone who wished to train himself could have access to those methods whereby he could render himself painless at will; even then, with that degree of proof, it would be possible to remain unconvinced that pain and its control had, or could have any larger significance.

If, however, we could find other parallel evidence of raw feeling being drawn off by the requirements of higher conscious activity, this case would be considerably strengthened. Such evidence, it would seem possible, might be found in physical pleasure. Pain and pleasure, agony and lust, are the two fundamental polar sensations which lie at an equally rudimentary level. Lust, however, is in some respects a better subject for this inquiry, because, though a less spectacular symptom of our still pent energy than is pain, physical passion is a more constant phenomenon. Pain may be dismissed as morbid or aberrant, but lust can be studied as normal and constant. We may ask: Have we here another and far more extensive example of the discharge of evolutionary energy? May we, by studying sex, hope to find how that energy might be channelled and, held from escape on more rudimentary levels, give rise to a higher degree of consciousness?

In the following chapter that inquiry is made.

CHAPTER III

THE UNUSED ENERGY REMAINING FOR FURTHER EVOLUTION

(B) —The Present and Future Significance of Sex

WE HAVE seen in the preceding chapter some evidence for supposing that the meaning of pain has been misapprehended. Owing to the assumption that man's evolution is over, it was taken for granted that any violent sensation could only be a notice to re-establish the *status quo*. If, however, our evolutionary growth is not over, then pains might be growing pains. The meaning of suffering might be a blind urgency rising from the depths of the organism, and pressing up to the threshold of consciousness. So the psyche is being summoned to open new channels for the still unexpended energy, and to prepare new moulds in which the flux of creative force may cast itself.

If this is the explanation of the anomaly of pain, we should be able to find evidence of the working of the same process in less exceptional manifestations of human sensation. What breaks out as a lightning flash in agony should be able to be detected discharging itself in a less violent and more constant manner. Many psycho-physiologists have noted that agony appears in many cases to be an intensification of feeling which in its lower registers is experienced as keenly pleasurable sensation. However that may be, the new theory of pain does suggest an answer to a question even older and of far more general interest than the problem of pain, and one over which there has been far more bitter controversy. That is the question of the significance of sex.

Here too, as in the question of pain, man's thought on this problem has gone through great change as his views about the universe altered. When his outlook was anthropomorphic he held violent physical pleasure to be, like violent physical pain, a thing sent by the gods. When his view of the gods was low, he believed that violent pleasure was the bait of a trap, as violent pain was the trap's teeth. When his theology became nobler, violent pleasure once again was classified under the same explanation as violent pain. Lust, just as much as anguish, was sent to try and test a man's self-control. God was peculiarly pleased with the pure, as with the patient, and virgins held much the same court precedence in heaven as martyrs. Similarly, when anthropomorphism yielded to mechanomorphism, the moral value attached to chastity faded *pari passu* with the moral value attached to suffering. It is true that Western Civilization has been slower in making up its mind about the uselessness of continence and the right to lust, than about the uselessness of pain and the right to analgesia. This may be due to the fact that effective physical anaesthetics were discovered at least a generation before reliable contraceptives, and that the progress of surgery increasingly exposed patients to sudden accesses of man-made suffering, while, owing to our natural secrecy about our sexual pleasures, most people were able to obtain considerable satisfaction while keeping up the appearance of practising temperance, if not of being total abstainers. Certainly, as with the discovery of chloroform and ether, the duty and value of physical agony began to lack advocates, so too, with the rapid improvement and knowledge of contraceptives, the duty and value of chastity was increasingly challenged. It may indeed be that we should never have heard of Freud and his psychological theories, that his name would be more minuscular and emotionless than the name of Janet, had it not been for the nameless inventors of unmentionable apparatus whereby the liberty which Freud preached could be practised without penalties.

Yet however slow our Western Civilization may have been in deciding that lust was to be explained in a similar way to pain, it has undoubtedly reached that conclusion. Most people who consider themselves up to date, are satisfied that as pain is a pathological symptom due to an obstruction of nerve impulse, so too, lust is a warning symptom of a tension and pressure which if denied relief, must cause damage. It is also clear that that conclusion rests on mechanomorphism. If reality is an immense machine, if evolution leads nowhere, and the emergence of mind is an anomalous accident which can only end in extinction, then the willing restraint of pleasure is as unnatural as the patient endurance of pain. It is not merely that man is denying himself a stimulant and sedative which blind chance has thrown in his way, a sedative which may ease him to endure the fact that understanding has shown him that life has no purpose. Lust, like pain, is a warning to restore a strained balance. By yielding to physical passion he is not merely taking a pleasure which only bigoted superstition would forbid; he is also performing an act of psycho-physical hygiene. Health being static, and not dynamic, we are called upon to avoid strain. There is no further development ahead. In staying as we are, consists the only real happiness. The end of life being physiological and not psychological, its goal being health and not understanding, there can be no reason for restraint. This undoubtedly is true, if there are no further ranges of awareness in front of us. If there are no new faculties to emerge and develop, then blind restraint may well be as dangerous as mortifying a limb or repressing an abscess.

There have, however, been throughout history a number of remarkable men and women who subscribed neither to the modificatory nor to the expressionistic theory of sex. They have not imagined either that there was a deity who was pleased with the degree to which his creatures could thwart their impulses, or that such impulses serve no purpose save to give momentary physical release. These are those mystics, and other psychological pioneers who have frequently left it on record that they considered inhibition to be a grave mistake, and that they only practised what appeared to onlookers to be restraints and austerities because they required the energy so reserved for other purposes. The Freudian answer to this assertion is that such types, owing to the intensity of their repressions, produce in themselves hallucinatory feelings. The mechanomorphic view of the psycho-physiology of sex can of course arrive at no other conclusion. The orgasm, being a purely physical expression, is comparable to the eliminatory processes of the body. A gorged condition requires release. The only significance of the sex act, as far as it is a conscious experience, lies in its power to de-tention the nervous system.

It is necessary to repeat that if mechanomorphism is true, Freud is right. We must stress this because any criticism of the popular physiological psychology is generally dismissed as springing from a disgust at all sensuousness (which is really repressed craving), and a morbid love of mortification. Alternatively, it must be stated, if mechanomorphism is an inaccurate picture of reality, then Freud's explanation is also inadequate. The Freudian position assumes, as is assumed by all mechanists, that man will never become a more advanced species than he now is. Any alterations can then be dismissed as morbid developments. They cannot be evolutionary mutations aiming at a higher consciousness. *Homo sapiens* is no more than the cleverist of animals, so clever that he may be too clever by half. He may be capable of using his overgrown intelligence, now that it has raised him above all the other animals, to cut away the branch out upon which he has climbed. The notions about his present latent capacities and his future possible developments are pathological day-dreams.

If this is the truth and the whole truth, love-making is, as the Romantics have maintained, the ultimate goal of life, the final activity of man. It is his culminating effort, after which, as the fairy story of the Romantic Age invariably concludes, he lives happy ever after, or as our more physiologically minded generation would say, in franker detail: the wave of life has swept out of him, and he is left empty, without urge or purpose, only desiring dreamless sleep. For, as Freud has pointed out, the orgasm is curiously akin to the epileptic seizure which obliterates consciousness and which has therefore been called "the little death, death's foretaste."

If, however, our evolution is not over, we need to consider sex from an alternative point of view, and when we do so, we discover that as a matter of fact our sex has many features which warn us against taking it to be a final expression. Indeed, it is as hard to force the facts of human sex into the Freudian explanation, as it is to force the facts of human pain into the explanation that pain is no more than a warning.

To-day there is common agreement among students of the subject that human sex shows three peculiar features which distinguish it sharply from animal sex. The first striking fact is that *homo sapiens* has lost his periodicity. He can be incessantly sexual, a doubtful privilege to which, as far as we know, the only other genus to be admitted is the large ape, the Orthograde Primate. The subconscious tides of life no longer direct that flow of his reproductive energy, so that it shall run as and when outer circumstances most favour generation. Secondly, not only is his sexual impulse untimed, it is undirected. Three facts establish this: the secondary sexual characteristics have usurped primacy of appeal over the central sex features. Even the grossest of men and women fall in love with a face or a foot, with hands or eyes. Next, the actual reproductive approach has been reversed, and finally the reproductive instinct has been so dissipated that not only has man no innate knowledge of the procedure of coitus but the impulse can turn back toward the same sex, or even spend itself in aimless auto-eroticism. The third peculiar feature about man is that he is not more lusty than the animals but he is twice as highly charged.

As has been noted, when reviewing our evolution, the enormous provocation which smell exercises in animals is with man largely eliminated. For the olfactory centre which is the dominant sense in nearly all other mammals is with man occluded by the growth over it of the Neopallium. The Neopallium, as we have seen, is a mental centre of the brain. Reflection therefore takes the place of impulse and the sense through which the reflective brain increasingly obtains its information is through sight, not through smell. This in itself is to shift interest from stimuli predominantly sexual to those only indirectly so. The change in the endocrine balance also throws its weight in this direction. Most mammals have a relatively large development of the supra-renal glands in comparison with their thyroids. The supra-renals are glands of sudden passion—the thyroid the gland drawn upon for persistent effort. In man this balance, set toward impulse and action in the sub-human, is set as distinctly toward reflection and understanding. Man is therefore not aligned as is an animal to take an excluding interest in specific sex. Further, when such interest is presented to him, it is presented in such a way (through sight) that not merely is immediate satisfaction, immediate discharge in action, not possible (and even not desired) but the crude impulse is alloyed with curiosity.

Yet, though his satisfaction is thwarted by these checks and distraction, so that he falls in love with an appearance only to find there is no comparable bodily satisfaction behind it, his charge of physical desire would appear to be wholly disproportionate to his reproductive needs. He has long been an animal which, through its supreme intelligence, was freed of many hazards to which all other beasts are exposed. He could hope to rear a larger number of his breed than any other species. Yet this creature, capable of a degree of foresight, of collective responsibility and of protracted parental care, beside which all other animals are hopelessly shiftless, is nevertheless endowed with a sex hormonic charge some ten times greater than that of a bull. At the same time this charge is set on a "hair trigger" release. An intensely irritable nervous system at the spine's base means that a very slight stimulant can detonate the mine.

If we consider this problem only from the physiological point of view, we are certainly confronted with a unique state of affairs. If man has no further development in front of him, if this huge charge cannot be transferred to some creative activity, we must ask: Is this over-development in the nervous system comparable to those local hypertrophies which occur in all species when general control vanishes and so their evolution ends in the production of non-viable types? Even at glandular level some indication of a possible purpose has come to light quite lately. For it is now known that the sex hormones cannot act unless the master gland, the pituitary, sends out a non-sex hormone to bring them into action. As the pituitary is a brain gland—situated just behind the root of the nose—and one in close touch with the higher mental activities, it would look as though this immense sexual charge need not be intended always and solely for reproduction but might, if a right impulse from a higher centre could be given,[1] be switched over to energize some new form of

1. See also page 185, Chapter X, where Laya Yoga and Vaijroli methods of sex control for enlarging consciousness are noticed.

development. Certainly as they stand to-day, the full facts of sex refuse to fall into any reactionary or mechanistic explanation, quite as much as do the facts known about pain.

All these anomalous facts fall, however, into place if we make the alternative hypothesis to Freud's, if we assume, as other evidence compels us to assume, that man's evolution does not necessarily finish at animal level, that he and he alone of all the animals, has a further, unprecedented development ahead of him, if he will use for that purpose the energy still stored within him. For, with what are we actually presented if we study our species without prejudice? Not a being who is blinded with excess of lust, but one who is on tenterhooks from nervous energy. Pent forces, not of instinctive animal appetite craving for specific satisfaction, but of neuronic impulse restlessly desiring it knows not what, fret and strain him. Indeed, that modern man's central problem is nervous hypertension which cannot yet find adequate expression and relief; that sex and the increase of sexual intensity and variety is only one symptom of our pent psychological energy, is shown by scores of symptoms peculiar to our species, and all increasing in variety and intensity since the beginning of the modern epoch. From frantic nationalism and the purposive cultivation of brutality (in the hope that in hardship and warfare there may be created an outer pressure sufficiently strong to counterbalance the inner), through organized athleticism, which calls for ever more stringent training, and ends in ever more frantic effort, to the sternutatory de-tentioning obtained by the snuff taking priest who is denied both violent exercise and sexual relief, we have a whole front of human experience seeking blindly how it may abort the embryonic power growing within it, and be delivered from and not give delivery to the new birth with which it travails.

Whatever his faults and failures, modern man is not seeking to be glutted but to be assuaged, released from an intolerable stress. The demands of undeveloped faculty rend at him, striving to find delivery in expression. He short-circuits the charge and finds a passing relief. He cannot hold out: he does not know the way to discharge this terrible craving in those intensities of focus that would give sufficient creative expression to fulfil that raw force. The fountain of life which in him should rise in a rainbow-creating jet, leaks away in a seepage of unrestrained, trivial eroticisms. Modern man's incessant sexuality is not bestial: rather it is a psychic haemorrhage. He bleeds himself constantly because he fears mental apoplexy if he can find no way of releasing his huge charge of nervous energy. Meanwhile the entertainment industry is largely devised with the express purpose of keeping his veins open; a hot bath in which he may die painlessly. The tragedy of our education is that to-day it tells us nothing of the methods whereby creative catharsis might be attained. This tragedy is all the more serious because at the very time that education has become completely ignorant of such knowledge, our species for lack of it is in danger of staggering down into extinction. For, without that knowledge, we must oscillate between the *tedium vitæ* of sex exhaustion and the frantic destructiveness of the Puritan repression which ends in the monomaniac militarists.

This association between sex and violence is still not generally recognized, and the few who do so generally fall on to one horn of the dilemma or the other—either that debauchery is better than militarism or vice versa. For a long while sociologically minded historians have been puzzled by the fact that militaristic dictators always interfere with the sexual freedom of their subjects, even forbidding behaviour which is private. "Why," it is asked, "do these tyrants set up unnecessary opposition against themselves? All they wish to do is to enjoy absolute power. What does it matter to them how their slaves spend their off-time? Is such interference worthwhile: may it not provoke unnecessary resentment? On the contrary would it not be Machiavellian policy to let these subjects debauch themselves? The one thing the tyrant can never countenance is a conscientious and critical interest in the State. The only person he has reason to dread is the man who is prepared to sacrifice his life to change the Government. Such a persistent interest and such an intense courage are least liable to appear among enervated people. Has it not lately been found in certain Latin American States that when mobs were gathering in the streets and demagogic orators seemed about to be able to raise a revolution, the tide was turned by just such tactics? The cinemas were ordered to be open free of charge and to show pornographic films hitherto censored. The public rushed to the easier excitement, the less risky pleasure. The orators were deserted and arrested without protest. The wronged populace was gloating on other delights than Liberty."

Yet the matter is not so simple as that or we should not have such able psychologists as the dictators, all of whose lives have depended on their power of understanding their peoples' temperament, always denying to their subjects non-political, sensual pleasures. True, dying autocracies such as the Russian Tsardom, and, on a pettier but not more squalid scale, some Latin-American adventurers may strive to distract the people from any interest in policy by leaving free to them pleasures which destroy purpose, and amusements that dissipate understanding. But no autocracy starts that way. Every tyranny, most of all the most modern, is an adventure, an adventure which calls its followers from pointless following of pleasure to attempt grandiose achievements. Such States are, and of that nature must be, always on the move, lunging, swelling, outbursting. But how are they to do this? Not by the actual muscular effort of a man whose chief physical development is his lungs. A dictatorial country snatches and holds the initiative just so long as the repressed energy of its rank and file is boiling to dominate its neighbours. That energy, as Dr. J. D. Unwin has pointed out in his *Sex and Culture* (Oxford University Press), springs from, or is an invariable concomitant of, sex repression or thwarting. The dictators know intuitively how their requirements may be met. They require not merely cannon fodder but fanatics. They obtain their fanatics from those who think that militarism is better than debauchery—if not for themselves—then for others.

That is the true gravity of our situation: If our evolution is over, if we have no further original outlet adequate to our enormous and fretting energy, then the only choice is slow degeneracy through sex addiction or a convulsive end through homicidal mania. We can only answer "Yes" to Matthew Arnold's desperate question, "Madman or Slave, must Man be one?" though our situation is even graver than the poet foresaw, because it is the slave who thought he would win security who is driven to suicide by the master to whom the slave sold his liberty for protection.

If, however, new faculties are ready to be born of us, then we require all our vital energy to give those faculties expression. Contrariwise, as we develop these interests and faculties, our raw energy is used up in this higher activity. Crude sensation, whether of pleasure or of pain, disappears. But that is not to say that insensitiveness or numbness is the result. Primitive sensuousness, the more violent it is, the more it stuns consciousness. Only when this dazing sensationalism is transcended, can consciousness experience sustained intensity of being. It is when interest is completely engrossed in an experience outside the sensory self that consciousness is most alive. That is why the artist, wholly lost in his work, and the mystic engrossed in his insight, have rightly been said to be in ecstasy—ecstasy which means the height of delighted experience and also the standing outside of oneself. Here is detachment which is not cold and distant but more in touch with reality than those, who are insulated by sensuousness, can ever be. This is also what the psychologist, watching the healthy and happy transfer of blind libido-energy into creative intensity, calls rightly sublimation. With these new facts about pain, sex and emergent faculty, we can describe this process by another term beside ecstasy and sublimation. We can say that this process indicates a possible ending of pain, a possible solving of the problem of sex, and also the possibility of a completely new step in evolution.

Yet a question remains to be answered, and it is a grave one. Granted that this great energy exists in us, waiting to be employed in the creation of a completely new faculty, giving a completely new insight; granted that this is the capital with which to build a new order; that with this power we may produce a new evolutionary social mutation, and widen the aperture of consciousness, can this happen of itself? True it may be, that a fresh birth of understanding is ready to be delivered. Will this delivery take place spontaneously? On the contrary, is there not increasing evidence that this cannot be? Is not our present distress and chaos, our psychotic, competitive and anarchic world, proof that we are failing to develop? All earlier evolutionary changes took place without the conscious resolve and attention of the animals in whom the mutation operated. To-day that cannot be. The impulse, the energy, the faculty may all be to hand. We may realize that a completely new advance is required of us. We may diagnose that the very restlessness which makes us alternate between despair and violence, is evidence of a vast energy only to be satisfied by an equally great enterprise. We may, even, have discovered that to-day there is emerging a faculty which can both absorb fruitfully that perilous force and can also give us apprehension of a new order in the place of our present chaos. Nevertheless, all this is insufficient unless we can discover a technique whereby we may release the energy along the channels by which it can inform the faculty and make this faculty part of the normal functioning of consciousness.

The question which this inquiry must now face, and this theory attempt to answer is therefore how we may cooperate with evolution, and realize our developmental possibilities. We shall be the better able to answer that question if we could discover more precisely what has been the trend of evolution as it ceased from the development of new organs and turned to the development of expansion of those senses which the organs provided with instruments. The outline of this specific extension of the senses during the period that man has been completely dominant in the world, i.e. since the rise of culture, is given in the succeeding chapter.

PART II

EVOLUTION THROUGH THE PSYCHE

CHAPTER IV

(A)—SENSORY EVOLUTION

The Historical Evidence for the Emergence of New Faculties

WE HAVE seen that there is reason to suppose that in man, and in him alone, there lies the possibility of further evolution: that this further stage must be through evolution of new faculties: that man is offered the possibility of emerging on to a new level of conscious being, as much above his present powers and apprehensions as they transcend an amphibian's: that the symptoms of this latent creative energy, pent within him, are the peculiar intensities and persistencies of both his pain and his lust. We have seen also that man's specific evolution, his development throughout history and indeed pre-history, has been psychological and not physiological. Therefore, now that evolution takes a further turn and is stepped up to a higher intensity, his mental evolution (which through the prehistoric periods was certainly subconscious and which through the historic periods has still in the main been unintentional and unapprehended) must henceforward become conscious. If we are to advance to the achievement and mastery of new faculties, we must do so deliberately, and, by a scientific technique, understand how to enlarge our apprehension. This is the task of psychology, the science of consciousness.

Unfortunately, when we come to study this, the favourite science of to-day, it gives us no help. Psychologists cannot teach us to control our minds, still less how to expand them. They themselves are not remarkable for their self-mastery, even less for their vision. Indeed, as one of the most eminent of our academic professors of mental science has remarked, "Psychologists generally are unhappier in their human relationships than the average man: they were driven to study the mind because they discovered by constant mistakes that they had no natural art for getting on with their fellows." It is clear, moreover, that the discovery of the lack of natural art was far from endowing them with a science. Such a state of affairs was, however, to be expected. Psychology comes last in a great series of the analytic sciences which, since the beginning of the modern world some four hundred years ago, have made their impressive attempt to bring all reality under the first principles of physics; to show, by analysis into their ultimate constituents, that everything obeys rudimentary laws of movement, and that a catabolic necessitarianism was the one principle expressing itself through these laws.

This principle works with a high degree of accuracy when we study the inorganic. Indeed, it was without exception until modern apparatus permitted perceptions far finer than our unaided senses permit. With the organic, with living processes, the principle did not work with the same unequivocal accuracy. Under careful observation it became clear that many processes of growth could be interpreted as easily as being anabolic as catabolic, while not a few appeared to be more probably teleological—the blind striving to an end, and the craving for a satisfaction. When we study consciousness, the mechanistic analysis breaks down. The hypothesis that the nature of mind can be understood by dissection of the brain has yielded such poor results after years of experiment that it can only now be maintained as a dogma.

Yet what is science to do? What other hypothesis is at hand and can offer a workable procedure? Granted that Kohler and Koffka have established the Gestalt theory—as there seems little doubt they and other researchers have: granted that the mind cannot be understood through atomic analysis, but works as an irreducible whole: granted that Goldschmidt is right, that the entire chromosome is another such gestalt, and so not only consciousness but the basic physical element of all living forms resists analysis; are we any further? A four-hundred-year-long attack which yielded spectacular initial successes has, when it closed in on the citadel, collapsed. The fundamental secret, which should close all inquiry and give perfect power of forecast, has escaped.

But if it does not yield to this method of analysis, what other method is left to us? This is the crucial question now confronting not merely psychology but all the sciences. Physics is equally embarrassed. It too has to ask, can analysis take us further into comprehending the universe around us, even in its simpler, inorganic manifestation? Yet if analysis has brought us to a standstill, what other methods offer any relief from the deadlock? We know that we are now faced with the problem of epistemology—of how our minds take in the universe around us, of what H. G. Wells called in the last century "the scepticism of the instrument," but how are we to solve that problem? It is, of course, as has been said, a question for psychology. Only a real science of the mind could answer whether and how consciousness could be enlarged so as to apprehend a larger range of reality, to see that reality more nearly as it is and so approach to an insight into the meaning of the whole and of the relationship of the individual consciousness to that whole. Yet the entire development of science shows that it has pursued a path which leads to the present deadlock, and in particular to the immobilization of psychology.

We have then to face the fact that analytic-necessitarian science, great as have been its material gains, has been on the wrong track, if its aim was, as its founders believed, to disclose at last a complete objective picture of reality. The analytic approach has yielded all it can. The assumption that the data of the senses yielded an accurate and complete picture of reality and that all that was needed was logical deduction from these premises and analysis of these experiences is now known to be inadequate. What is needed is what Radhakrishnan has called the integral approach. How is that to be obtained? It seems clear that if we are ever to have a true science of the Mind (and thence with that science come to a new and enlarged insight into reality), we have to discover a method alternative to, and very unlike, the analytic method of inquiry. How can we invent such a method? Granted that, as Bacon said at the beginning of analytic science "I reject the Syllogism," we have to say "I reject Analysis," on what are we to build? It seems clear we have to search the past. If we are to advance we must cast back. We must go behind the analytic epoch and seek before that time for processes which might yield, at least, hints as to how we might achieve integral thought leading to a further extension of consciousness.

That human thought, when it faces failure in its leading stem, may "sprout below the graft" is not a notion as shocking to us as it would have been to our grandparents. History has taught us that minds in the past were as subtle and as forceful as ours. The difference between us is the different subjects to which they and we attended. More lately history has also shown that when a great culture exhausts the dominant line of its development, as Egypt did with Akhnaton's reform, when advance wrenched up its own roots; and as Greece did likewise in its turn; then, as we shall see in Chapters VI and VIII, the truly radical advance is made by returning to methods which had been impatiently discarded as superstitious. By finding the psychological equivalent of the traditional mythos, the new thinkers in Egypt and in Greece transmuted what had degenerated into an anchylosed art into a contemporary science. Anthropology has even more lately added, that peoples who have seemed incompetent because they lacked that engrossment with analysis which has given us our physical powers and our economic complexity, have in compensation certain mental powers and methods of attention which we have let lapse.

It would seem then that we have reached a stage in our evolution when under the terms and restrictions of our present apprehension we have gone as far as we can. Confining ourselves as we have since Galileo, to measurement of such movements as our present consciousness can perceive, we have disregarded as unreal such vague intimations and inhibitions as are experienced by backward people who have not attended exclusively to outer phenomena but have been still aware of their inner nature. Hence we find ourselves with a sanctionless morality and a senseless universe.

Now that we realize that we have restricted our possibilities of observation and that further ranges of reality undoubtedly lie beyond our present senses' apprehension, it is clear that what is required of us is that we seek an extension of apprehension, a broadening of consciousness which will bring new data, new experiences into our range of knowledge. It seems increasingly clear that the great adventure in physical knowledge which began with Galileo was also one of those specializations which, if not balanced by an equal advance in power of comprehensive apprehension, must lead to an increasing restriction of interest, awareness and understanding. First, there is a complete disregard of all but physical changes in the outer world and then even the interest in these changes is further confined by the ruling out of the secondary characteristics of phenomena such as colour and sound, leaving only "extension and mass."

This specialization must be corrected by recovering once again a generalized apprehension which does not reject any experience—be it colour or the sense of the self—whereby the organism lives. For the organism which begins to dismiss actual experiences because these disturb the simplicity of its explanations is headed for that ever-narrowing awareness which ends in presenting itself with a world in which it cannot live—a world not only without sanction but without meaning.

Before, then, we can hope to attain a science for the dilation of consciousness, we must search in our pre-scientific past and among surviving pre-scientific peoples, to see whether among them we may find the seeds of such processes. And we may pursue such a search with a patient optimism. We may be optimistic: there is much material at hand. When we view history psychologically, and non-scientific contemporary culture anthropologically, it becomes clear that we must be patient. For, however "viable" those seeds, they will need very careful rearing if they are ever to give rise to a science, a true psychology. To change the simile: Granted that if we neglect knowing what can be recovered of the mental techniques, of the past, and what can be told us by surviving practitioners of such techniques, we have no crystals from which to make the lenses we need, we have to remember that not only were most of the psychological methods of the past rudimentary (and that they remain rudimentary and may even have degenerated among the present inheritors of such methods), but also that even where such methods were exact and self-conscious, they were devised to fit a consciousness of different "focal length" from ours to-day. Though we may learn the principles of lens-grinding and focusing from them, we cannot reconstruct and use their actual spectacles. There is progress—inevitable progress, in the sense that each epoch is more advanced (toward larger life or more irrevocable death) than that before it. We are called on not to return to the past, but, with the sharp-angled spiral of ascent, to recover and translate in scientific terms (as our intenser consciousness demands) the intuitive knowledge which the past knew only as an art, and which our immediate predecessors dismissed therefore as untrue.

This inquiry must divide into two parts. This question of the evolution of consciousness is still so unfamiliar to nearly all of us, whether educated or uneducated, that we must first advance evidence of its taking place as part of the main evolutionary process. We must be able to show that man's consciousness has shifted and grown at first spontaneously, and that there is clear evidence that where there is no question of self-deception or vagueness of terms, such extension of faculty is both possible and has taken place. When we consider the problem in this way we discover that not merely is there unmistakable evidence of this, but that the "bud" from which this development has sprung, may be found at sub-human level. As the great apes share with man his equivocal freedom from instinct and sexual periodicity, so too they have undergone at least the first stage of that psychophysical development which finally can lead to enlarged, supra-animal consciousness.

The development which leads to the threshold of superconsciousness, after which a science is necessary if further advance is to be achieved, divides into three stages. First, there has been a steady shift from the shorter ranged, more strongly emotional senses (of which scent is the strongest) to the longer ranged and more speculative senses. This, as we have seen in the chapter on sex, accounts in part for our sexual anomalousness. The studies of the late Sir Grafton Elliot Smith have shown that even in the brains of primates, as they approach the human type, there is anatomical evidence for that regression of the olfactory centre as the neopallium growth comes forward, which neopallium, as we have seen, is one of the centres of high intelligence and finally in man practically occludes the older nerve centre. It is equally clear that in the primate, sensory evolution, sight is exchanged for smell.

Secondly, sight having become in man the dominant sense to which all the others are subordinate, the range of sight is steadily increased. It is possible in the growth of pictorial art to trace how, as man's freedom from immediate cares grew greater, so he literally extended his perspective. Primitive art is solely concerned with close-ups. Then, to add to the importance of the foreground, figures and incidents are worked into a middle distance. The interest in true perspective is significantly a modem development. Is there any reference in a classical author before Cassiodorus, to someone going out of his way to gain a view? We know that not until Uccello are Western artists interested in perspective, and that then the discovery appears so revolutionary and so fascinating that Uccello's wife cannot get him to come to bed. Even then the advance is gradual, and it is not until the Baroque has superseded the true Renaissance and the abstract speculation of pure science has begun, that a landscape art in which the view is on a scale to dwarf all human figures, begins to engross the principal attention of painters and patrons.

It would seem that the same slow expansion of attention marks the development of Chinese painting, and that it is a final development when extensive landscapes become the main interest of artists.

Thirdly, colour develops in the same way. Dissection of the eyes of apes and monkeys seems to show that they are provided with, as well as the retinal rods, the cones on which the apprehension of colour is said to depend. Experiments in testing them with coloured wools also seem to show that they can probably discriminate between the primary colours if they wish. The fact remains that this biologically unnecessary attention is not employed. Colour tests on young children seem to show that the child attends first to red and last to blue—following in the growth of his taste the shortening of the wavelength of light. In this the child would seem to recapitulate the expanding experience of the race. We know that paleolithic man, though a consummate draughtsman, only employed as his colours red, yellow and black, the Bushman artists worked under the same colour restriction, and all primitive races to-day appear to have a partiality for red and an indifferent apprehension of the other colours.

The Nepalese did not perceive blue until their attention was called to it by Europeans, when they called it blu-blu, but still it remains doubtful whether they perceive it as a colour or only, as the colour-blind, as a difference in tone. Rivers pointed out that the Fellahin had no clear words for blue, and that the name, the "Blue Nile" really means the Dark Nile; and that they confuse brown wools with blue.

In our own ancestry also, the uses of colour words show how rudimentary was our colour sense until comparatively lately. In Welsh, the word for the blue or white of the sky, the background, of field or wood or hillside, is *glas*—a neutral tinted word for the transparency of the distance, and which we have therefore taken over as the word for the most transparent and colourless of materials. Alternatively, the word for colour in Russian also means red.

In Homeric Greek the same rudimentary colour vocabulary is evident, Homer having half a dozen words for red, but no clear terms for blue and green. The eyes of his generation seem to have picked out the slight tinge of red in the purple of the Mediterranean, and to have been blind to the blue—hence his strange but constant epithet for the sea, "wine-dark." The painting of the deities' hair blue and calling such locks hyacinthine also seems due to a confusion in their colour sense. As most blues appear white in an ordinary photograph taken without chromatically corrected negatives, it is possible that the Greeks thought they were making all their gods platinum blonds. Nor did the limitation disappear even at the high noon of Greek culture. Their pottery, in spite of the increasing mastery of form, remained black and red. Blue and green glazes, though easy to make, are never employed. Far more remarkable than this, however, is the fact that such an apparently modern genius as Aristotle declares that the rainbow has only three colours. To the red and the yellow he can add the green, but even he was evidently unable to perceive the violet.

We can then conclude that in man's expansion of attention first the farthest ranged of all the senses becomes dominant. Then this sense increasingly takes pleasure in attending to distance, to experiences which cannot be enjoyed by any of the other senses. Finally this sense, having established not merely its predominance, but its independence, develops a new faculty, a faculty of little physical use, the appreciation of colour.

So much for the clear historical and contemporary evidence that modern man's senses have grown, that there is psychological evolution, and that that evolution is in the direction which leads man to the apprehension of a larger world. Man's sight becomes his predominant sense; using it he becomes increasingly interested in the distance, in ranges which are outside the limit of his appetites, and sight itself is continually enriched by the growth of the colour sense.

This, however, is only the first stage of our inquiry. We have found evidence that man's apprehension of the physical world has grown immensely in richness. Compared with the animal's outlook he may be said to be possessed of another quality of consciousness. Yet this is only an enrichment of the same material world which canalizes and limits all bodily behaviour, whether of man or of beast. There is no new apprehension but a shifting of the accent in the use of the senses, and a heightening of the predominant sense, the sense of sight. In short, all we have considered so far is hyperæsthesia. True, the tendency of such hyperæsthesia, the evolutionary direction which the development of the senses has taken, does suggest that mankind has been unconsciously feeling his way toward the wall of animal limitation, and that there might be within our race not merely a desire but a capacity to transcend those limitations. Can we hope to surpass our circumstances and obtain an insight which would show us not merely the physical world (wide instead of cramped, and coloured instead of plain), but another order of things, in which violence would not have to be the sanction of morality and values would not be alien luxuries, only to be enjoyed under a protection which destroys their flavour? As the range of our ordinary senses might be compared to walking, and hyperæsthesia compared to running, so in this simile, this radical extension of faculty would be comparable to flying. It is the power to proceed in a completely different and superior medium.

CHAPTER V

(B)—PSYCHICAL EVOLUTION

The Evidence for the Evolution of Consciousness

THE EVIDENCE already put forward in this inquiry suggests three things. First, that evolution continued latest in the higher primates and especially and finally in their most advanced species, *homo sapiens.* Secondly, that this evolution has ceased to be physical and become psycho-physical—new organs have not been developed but the farthest ranged sense has been advanced at the expense of the shorter, and instinct,[1] or random trial and error has been discarded for experiment. Thirdly, that in men, this now paramount sense of sight has also been enriched by the addition of a constantly widening colour sense. Further, detailed study of the power of vision in apes and monkeys seems to show that in these animals the apparatus of the eye—the equipment of cones in the retina—is organically adequate to apprehend colour, while experiments on the behaviour of apes would appear to indicate that if they wish they can distinguish between a number of colours.

These facts taken together suggest that the latest evolutionary development is one of attention, in other words psychical and not physical, an evolution not of organ but of consciousness.

Man, we may then assume, is an animal which is still evolving. His senses are becoming finer and are showing him a larger world. That world, we must however repeat, is no more than the animal world, a world which is meaningless to intelligence and whose only significance is that it contains objects which temporarily assuage appetite. The fact that we have come into this expansion and enrichment of

1. If indeed Instinct was ever highly developed in any of the unspecialized successive species from which man has been evolved. If Instinct is "fossilized trial and error" then the unspecialized types would never have developed true instincts for these are the psychological concomitants of physical specializations which are symptoms of evolutionary failure.

our sense's range makes us therefore ask all the more urgently: Is this the end?

By itself, hyperæsthesia is not inevitably a reassuring symptom, for if we are, with all its aid, seeing no more than the old animal world with a new intensity of focus and a new irrelevance of colour, we may just as well ask whether this is not evidence of our decadence and doom, as whether it is proof of evolution and progress. If our hyperæsthesia serves only to bring out in sharper outline the fact that we are inextricably surrounded by an iron world of blind mechanism, then that we see the bars and teeth of our contracting trap iridescent with ever-fresh colours is, if it is proof of anything, more likely to be evidence that our fate has driven us insane, that we are morbid species, than that we are growing to a stature which will permit us to see our way out.

Yet this fact that evolution has continued in us and that that evolution has shifted over from the physical (from the evolution of organs) to the psychical (to the evolution of higher awareness and attention) gives us grounds to ask whether there might not be evidence of even further evolution, whether beyond hyperæsthesia there may be found traces of radically new ways of apprehending, whether, as physical evolution has become in our animal stock, psycho-physical, in ourselves, in specific mankind, psycho-physical evolution might not in its turn be yielding to purely psychical advance.

In this branch of inquiry it will be found that we are not without evidence. We are, however, faced with a difficulty in this second and more important part of our inquiry, which did not confront us in the first. Whereas, sense development has been carried on in the cultural and psychological heredity to which we ourselves belong, so that the colour sense of mechanistically minded man is probably keener than most savages', integral apprehension has been increasingly discouraged and neglected by our practical minds. Uncritically assured that we knew what our entire natures needed, and assuming, as unreflectively, that the basic nature and meaning of external things is to yield us whatever we desired, we have become increasingly contemptuous and even unaware of any method of apprehension other than the analytic.

We cannot, therefore, expect to find this faculty—integral apprehension—present in us. If it does appear sporadically it can only do so in individuals who, either through their inadequacy to understand or to acquire our standards, are educationally so low as to be beneath our serious attention. This fact, that such who still appear among us with the faculty of wider apprehension, are probably incapable of conveying what they see, or of sifting fancy from apprehension, compels us to seek for its appearance and employment in the records of the pre-scientific epoch and among such peoples who today have not become subject to analytic thought. Our pride in our achievement in mastering our environment still makes it difficult for most of us to allow that people who lie behind us in time could have been ahead of us in apprehension, and that an oriental or even a savage who lacks our physics may be beyond us in empirical psychology. Yet if we are to be exact rather than resentful, we see that we are not called on to make a sacrifice of our material progress, still less of our scientific integrity, to the incompetent and inefficient. We are not asked to go back to savagery and abandon our economic equipment. We are not even asked to take over the psychology of the past, nor that of the contemporary savage. What we are invited to do is to seek among such for certain branches of developing inquiry which our immediate ancestors neglected when the analytic mechanistic outlook began to spread from physics and wrongly to invade psychology. Further, when we have found such branches, we have to develop them.

For fear this inquiry may still be considered beneath our dignity, it may be apposite to quote the opinion of a great geneticist on the disappearance of abnormal faculties in the modern epoch. William Bateson, who did so much to recover the work of Mendel, was interested in a peculiar historical mystery about which there has been much speculation, the witch trials which were carried on throughout Europe from the beginning of the fifteenth to the end of the seventeenth centuries. The evidence is carefully and copiously recorded, the procedures of the courts were frequently just and patient; evidence was sifted, witnesses were examined. How did the judges deceive themselves and condemn thousands who, whatever their faults, could not have been guilty of the charges for which they were executed? Bateson was persuaded, as have been nearly all careful inquirers, that there was a basis of fact underlying the accusations and—even more remarkable—the confessions. He conceived that when the witch-hunting was at its height there were present in Europe a number of stocks which carried as one of their hereditary factors some sort or sorts of hypernormal faculty. These stocks the executions exterminated, except in the "Celtic Fringe," where the Inquisition, and its successors, the Puritan witch-finders, did not operate. This explanation of a historical mystery at least throws light on many otherwise anomalous facts. It would, for example, account for the fact that second sight and other abnormal faculties have continued to appear with sufficient frequency among Celtic stocks, that no one who has lived among such could doubt the existence of such faculties, while among the peoples exposed to the centuries of witch persecution, the survival of such faculties is so rare and so weak that any rumour of them can be dismissed as superstition. To follow this hypothesis: the very success of the witch-finders led to their own dismissal. By the end of the seventeenth century it is possible for Selden's view to be increasingly accepted among educated and influential men. The belief in the fact of witchcraft can be laughed out of court because no witch has survived to demonstrate anomalous capacities. Such an unexpected success on the part of religion did much to undermine its own claims, for the enforced disappearance of nearly all who had any apprehension other than through the common senses (which saw nothing but the material world) made mechanomorphism possible. So a peculiar convergence[2] permitted humanity to reach its present pass. The destruction by the Church of the rare stocks who had exceptional insights chimed with the extension of mechanism, from a hypothesis which covered the visible facts of physics to a dogma which was extended to embrace the whole of reality. First the buds, out of which extended faculty might be derived, were destroyed, and then a system, which maintained

2. For the connexion between Orthodoxy's success and Materialism's rise see end of Chap. XI.

that such a faculty was impossible, was constructed.

It may be claimed that the Church, though it persecuted the witch, fostered the saint, and the saint had the same supernatural capacity. It is therefore necessary to point out that throughout the same centuries in which Ecclesiasticism was busy exterminating the witches, it was also repressing the mystics because their apprehensions did not confirm the Church's dogmas. The victory of the Jesuits is, as we shall see (Chapter IX), a victory of material representationalism (a representationalism which showed another world as concrete as this) over a dawning realization that here and now the human spirit is surrounded and interpenetrated by a reality which a new faculty is needed to grasp.

To-day, however, that procedure has been completely reversed. The long cessation of persecution has permitted the few stocks which escaped extermination once again to become comparatively numerous. Hence abnormal faculties are not now so rare that an unbiased and careful researcher can fail to come across them, if he is patient enough to work through a certain amount of fraud and unconscious deception. At the same time, while such evidence is no longer difficult to find, mechanomorphism no longer covers completely the realm of physics. Its claim to account for the facts of growth and of consciousness has had to be abandoned. Our minds are therefore free to accept the evidence on which a larger, more adequate cosmology can be built. The evidence, however, up to the present seems to most inquirers sufficient only to upset mechanomorphism, not to put another cosmology in its place. For that we, no doubt, need to realize where further information may be obtained. Yet it should now be clear both as to where we must search for rudiments of emergent faculty and as to what methods must be employed if that faculty is not to remain rudimentary but be extended until it is able to be employed intentionally, until it can serve its full evolutionary purpose, a complete expansion of consciousness. To discover what techniques humanity has employed in the past to dilate the aperture of consciousness, we must consult history.

Unfortunately the history of such exploration and experimentation is the most obscure of all mankind's records. The history of ideas is the most important history, but it is both scanty and cryptic. This is so for two reasons—reasons which have to do with historic man's evolution. In the first place, as long as man has tolerable access to his subconscious mind, he does not care, indeed he may be unable to be critical and self-conscious about his method of approach. The well-known rhyme about the centipede illustrates that stage and the dangers which man at that level senses to lie in self-consciousness. This shrinking from understanding one's methods does not, however, save mankind. So, in the second place: once self-consciousness has arisen, this fissured foreconscious (at least among those who successfully survive, and history is written by survivors), is at first not concerned with what it has lost, but only with what it has gained, the power to break the rules, shatter the tabus, and loot the sacred reserves. This is the "Heroic Age," the epoch of incessant destructive activity. It breaks down the last frail ruinous bridges still linking man with his past.[3] Later, the new individual, when action is over and death at hand, is now so sundered from the deeper consciousness which united him with the eternal life of the race that he can only think of being remembered by the saga of his deeds. Interest in the development of thought has therefore no place, even among the chronicles. Even in words, only action, violent action, is to be recorded. Hence the only early history is Drum and Trumpet History,[4] the *acta* of men whose deeds are almost all destructive and whose thoughts are sporadic and despairing.

The history of ideas, of how man actually understood and developed his technique of understanding (for his technique has to increase as his self-consciousness increases) is therefore unrecorded; in the first stages, because man shrank from defining his methods and in the succeeding period because there were no methods to record. The Hero has destroyed the old psychological way of life, and history is confined to saga-justification of this destructiveness.

---

3. See Peake: *Bronze Age and the Celtic World*. No hero could trace his ancestry beyond two generations, and so escaped out of the heraldic disgrace by calling in a divine misalliance, claiming descent from an immortal.
4. See Chadwick's *Heroic Age*.

Finally, as we have seen, the passion to exploit and destroy leaves man with no conviction but that the world is his oyster, that the universe has no meaning save to be the means to his physical satisfaction. To employ an old psychological classification—the Extrovert, unreflectively engrossed in the apparent world, has by thoughtless action, created a state of affairs in which only extroverts can survive. The Introvert, the man who could forward evolution by exploring and enlarging consciousness, is exterminated. Hence man is cut off from any union with, or insight into, reality. Hence he creates the analytic, mechanistic science which to-day gives him infinite means and no ends, unlimited powers and no sanctions. At length he can only perceive with general despair, what at the beginning he wished to believe from private greed—a world which is completely a-moral and yields its unlimited resources to those least trammelled by tabus.

Nevertheless, even such history as we have shows ample evidence of man's constant preoccupation with the problem of his fissured nature and his subconscious mind. Even saga history, the first chronicles, the records of the Heroic Age of Violence show that in that extroverted age, men still were acutely concerned with the question whether the practical, foreconscious mind, using the animal senses, gave them an adequate picture of reality. Even our first and most mechanistically-minded anthropologists recognized, and had to recognize, what a vast part religion has played in the lives even of the most brutal men of action (indeed in all but the most modern men's),[5] and that the basis and force of the religious conviction was the experience of the subconscious, generally, but not always, through dreams. Now that we have evidence that dreams not only give proof of a deep of the mind, following an undertow of thought which seldom rises to the surface, but further, that on occasions these profound currents flow out and intermingle with tides beyond the individual's limits, we can hope to trace, in the efforts man made

5. And should not this be modified when it is maintained that Adolf Hitler, the latest of realists, before whose power the old-fashioned practical men have to bow, regularly consults his astrologer, so that his greatest diplomatic victory was won because he was told the stars would sustain him in any venture undertaken on September 27, 1938. (See *New York Times*.)

to command these currents, some outline of his psychical evolution.

We have seen that such history as men chose to record disregards the theme with which we feel it should have been concerned: the evolution of consciousness. When man had a psychological technique of keeping in touch with the subconscious, he at first felt that the process or knack needed no record; then as the process grew less deft he dreaded to inquire too definitely, feeling that if he tried to regain a firm hold on it, it might vanish altogether. If he made notes, he kept them away from common knowledge. So we have no direct record of how he altered and made more precise these methods as his self-consciousness grew, of how man devised an ever more intentional technique, a technique required because this growth of self-consciousness caused the subconscious to sink ever deeper from his grasp, a technique made possible because self-consciousness permits man to devise scientific, objective methods toward himself. As we shall see later when considering the oracular technique, as used at Dodona, it may be that when the mind has reached an anthropomorphic outlook it can only express and keep in touch with the subconscious levels through anthropomorphic phraseology and procedure. So petitionary prayer would be the necessary medium or glove through which an individualized consciousness could make contact with the consciousness which is not broken up into individualities.

At the same time, although this essential theme is neglected, in recorded history, we can, with our present anthropological and psychological knowledge, detect what developments were going on in man's mind through history, and were giving rise to those actions which chroniclers recorded. We can treat the events of history as symptoms of this basic growth and so detect the path of its development. Anthropology confirms history that there are three main stages in the human mental development which has culminated in modern, industrialized man. First there was the pre-chronicle stage when union was preserved with the subconscious and so with the para-individual life by ritual by corporate dances. This is what may be called the kinæsthetic rudiment of psychology. This technique is, however, not advanced at a sufficient pace to keep up with the intensification of individuality, or perhaps we should say, the intensification of self-consciousness has to abandon such an intuitive method for one more deliberate, specific and intentional. Whether a break could have been avoided or no, we cannot say. That such a break did occur is clear. The Heroic Age, an age of violence which dismisses old balance and restraints as timid superstition, shatters the old stable pattern of living. After a short phase of almost intolerable disorder, an age in which humanity might have disappeared, had men not been living mainly in segregated societies such as the City States, the militarists themselves wish to impose order and attempt to salvage the old system. They cannot, however, do so. They can only set up unreal, anthropomorphic deities whose codes have to be enforced by the police power of the armed king. These deities soon become discredited.

It is then that we enter the third and final phase in our psychological history. Complete individualism appears. Everyone is now a private person. This conception gives rise to three things: scientific materialism, inorganic democracy, and industrialism. It is difficult to see whether there is any further possible development along this line of descent. A morality to hold society together is attempted by the doctrine of mutual self-interest. Such a morality is, however, insufficient to give to inorganic democracies the necessary social cohesion. Hence we have those frank reversions to the Heroic Age—the Dictatorships. They win against the Democracies (as happens, we now know, in most of the vital struggles decided by Natural Selection), not so much through direct conflict, but by drawing off the enthusiasm of the active to themselves and away from liberty.[6] They offer a fuller life; for a false conclusion satisfies a man better than none, and he will prefer a purpose, however dangerous and delusory, to a life which is free but without any aim greater than private satisfaction.

Now that we have this fundamental outline of modern man's development, we can place in that framework in some order, the

6. Hitler's forces are welcomed, if not by the majority of the populations he invades, at least by a minority so active that it can dominate the situation and open the gates to him. He has a majority of force, though not of numbers.

successive psychological attempts which men have made to keep contact with their eternal life through the subconscious. We can say that when the tribal communism through the dance ecstasy is shattered, the priest-king leader is slain and the hero brigand with his band of armed looters takes his place, a revolution has occurred, not merely in men's affairs but in their minds. It will henceforward become far harder to cross the threshold of the mind than it was before this cataclysm, not only because men will, be pessimistic, doubting that there be any just order in the universe, and believing blind chance is all, physical violence alone decisive, but also because their own minds will be torn and their attention increasingly fixed on the outer world. They will not be able to escape noticing and dwelling upon those happenings which confirm the belief that blind fate rules, and they will fail to attend to hints of any other factor at work.

It is at this time, therefore, that we should expect to find thoughtful men resorting to drugs in order to dilate consciousness. There is some evidence to show that even those typical heroes, the Homeric protagonists, were not able to find in animal activity sufficient satisfaction to make life endurable. (See further, Chapter VIII.) There are passages which describe them not merely as regularly besotting themselves with alcohol but also when, unable to endure grief at bereavement (for death is completely abominable to the hero) taking nepenthe—a powerful narcotic, perhaps hashish (Indian hemp) or perhaps opium. Distilling seems to have been discovered at least in the Middle Kingdom in Egypt and the use of alcohol as a religious means we know had a vogue which culminated in the Dionysiac enthusiasm, reaching Greece before the sixth century B.C. (See Chapter VIII.) No doubt Indian hemp and opium proved better dream generators, but as they are both highly toxic they had to be abandoned. They might dilate consciousness but they certainly dissipated it.

There lay the problem, and it seemed almost insoluble. It was possible to throw the mind out of its lower animal mesh: it seemed almost impossible to get it into a higher gear. To blot out for a moment the harsh imprisoning appearance of things was all too easy. To open a window on to a higher co-ordinating apprehension baffled all endeavours. The explorer found himself thrust back into the strait confines of the material world. Those who had witnessed his effort only carried away as evidence, glossolalia—incomprehensible babblings—and he himself, if he brought back any recollection, carried away only an even more incomprehensible euphoria, a mysterious but quite unevidential conviction of a larger experience.[7] This phase we may call Shamanistic, and at this place it is advisable to remind ourselves that man is attempting to obey a profound but obscure urge in his being, in fact, that he is striving at one and the same time to avoid the fatal tendency to specialization and individualizing analysis (which leads to the Heroic Age and militaristic violence and then to materialism and anarchy) and to attain integral thought and insight. He is aware of a tendency in himself for his consciousness both to expand and also to define. He feels that there are forces pent inside him which, if he could find adequate expression for them, would carry him to a new outlook on the world and new union with reality.

So these early researchers at the Shamanistic level know only that they must do two things. In the first place they must shun the ordinary ways of men. Man has lost his "open vision," to use an early Hebrew phrase for integral apprehension. War and violence, like a widening fissure, increasingly threaten to engulf the race if it cannot find another route. Greed, lust, and fear grow to be the only realities among men, and the universe appears to be completely indifferent. The true racial pioneer must strike out, and away by himself find the track which leads to life and not to death. In the second place he must find methods which will reunite his own dividing consciousness. Shamanism fails because this reunion is almost always a regression. Instead of raising both sides of the mind to the same level and making an integral thought equal in intensity to that which analytic thought had attained, instead of bringing a widened field of consciousness into clear, critical awareness, the Shaman retreats to a more rudimentary psychic level. Instead of throwing a bridge across

7. e.g. The typical gnomism brought back from an anæsthetic experience which seemed completely revelatory but seems to us hardly to yield a solution of all Life's problems: "The Universe has no Opposite."

the gap at the point which the evolution of consciousness had actually reached, he sank (and still to-day sinks) into his psychological past until he finds himself at that rudimentary condition and frame of mind at which the fissure had yet to open.

The Shamanistic phase (a phase distinctive because in it the dance, which was once corporate, is now done by the Shaman alone) is succeeded by the Fakiristic stage. The psychic pioneer, finding that he had increasingly to separate himself from the society and circumstances of his fellows, would also find himself almost inevitably confronted by wild nature. Denatured Deserts and National Safety Parks are very late inventions. Between the extremities of human congestion and inhuman desolation there was no middle term until men became rich enough to prefer and to be able to choose Romance rather than Reality. So, enduring extremities of hardship, the Fakir would discover, in that condition of strain, that he found himself not only freed from the pressure of social suggestion and of physical fear, but filled with a strange elation. Hence he would begin to believe that in the endurance of exposure, privation and pain, true vision, real expansion of consciousness, was attainable. This, however, did not prove to be so. The prophet-fakir might become an awe-inspiring character who could exercise considerable social power through the prestige which his fearlessness of the unknown, his stoical restraint, his athletic energy and his moral courage would command. He would, however, generally fail to attain to consistent seership because his method of dilation would be too rudimentary to yield steady results. After an immense effort, as we shall see is illustrated by the conduct of Elijah, the typical Hebrew fakir-prophet, he would collapse, falling into complete discouragement. This phase of psychic pioneering must therefore pass, and beyond the Fakiristic stage there would seem to lie the Yogic.

This triple series in psychic development seems to indicate the true development of mankind's thought and practice as he has attempted to achieve the evolution of his consciousness. For the Yogic stage is that at which man no longer attempts, by deliberate violence to himself, or by suffering it at the hands of nature, to enlarge his awareness. Intensification of conscious attention, deliberate detachment toward himself begins to show him that by concentration of the mind he can attain the goal he is seeking. He realizes that neither pain nor physical shock is the force which he is seeking to give him the requisite energy to make completely new psychic faculties function. Pain and pleasure, strain and relaxation are simply symptoms, indicators of the intensity of the unused primal energy still residing in him for further evolutionary use, if he cares to employ it. At the Yogic stage, restraint is, therefore, still employed, but not mortification. Restraint is necessary, not because pleasure is in itself weakening or pain stimulating, but because the basic energies may escape through these sluices, instead of turning the dynamos of higher faculties.

With the Yogic stage we arrive at methods which are no longer accidental or empirical, but which show a scientific attempt to attain precision of action and accuracy of record. The aim and means are clearly perceived. Such practitioneers at last realize what it is that man's evolution now drives him to seek: an energy which will be strong enough to drive him through the manhole of the steadily sinking submarine—the downward drift to specialization—in which we are all imprisoned, up again to the surface. To change the metaphor; if we will attain the new level of awareness we must be prepared both to fly our plane with the utmost skill and also to throw over everything which may now be useless weight. We still contain the essential energy, sufficient to permit us to continue our advance and to escape being sucked down into automatism and death, but we must consciously release it. For the existence of that energy, the early pages of this book show some evidence. Later we have to see how it may be controlled. Meanwhile in the pages immediately following, we must, in order not to neglect any methods, however rudimentary, which might help us, give examples from the history of our own social heredity, of the three phases of psychic evolution: the Shamanistic, the Fakiristic, and the Yogic.

In tracing these three main stages in our psychological evolution, the examples given here are taken from Egyptian, Hebrew and Greek history. No doubt if we had a full history of Asia and especially of that source-land of religions, India, we should be able to find far fuller confirmatory examples than in our own social heredity, which is rooted in the Levant. Yet there are two good reasons for being content with Levantine sources. In the first place, Sumer, Egypt, Palestine and Greece gave rise to the ideas which have shaped our development. Secondly, that development, although it has become gravely unbalanced in the last four hundred years, is one which was partially correct. Our ways of thinking are worth studying not merely because they are ours and can be dated but because they have a contribution to make to the thought of mankind as a whole. The fault of our analytic thought is not in the thoroughness of its analysis, the mistake of our physical development is not in the intensity of our consciousness. On the contrary both of these developments have been necessary and should be beneficial. What has made these growths dangerous is that they have not been paralleled and balanced by a proportionate development of integral thought and of widened awareness, dilated consciousness. Indeed, it seems quite clear that though the analytic West must gain new areas of study from the contemplative East, the East must also gain the full technique of accuracy, definition and experiment from the West. With widened range we must preserve the highest clarity of focus.

CHAPTER VI

THE HISTORICAL EVIDENCE FOR THE EVOLUTION OF CONSCIOUSNESS

I—Sumer and Egypt

TO START then with Sumer and Egypt: Dr. Frankfurt, who speaks with first-hand authoritative knowledge of these two large fields of dawn-history, believes that certain generalization can now be made.

In the first place, excavation in Sumer convinces him that there has now been uncovered in the lowest levels of the earliest cities, the pattern, imprint and precipitation of a way of life which shows what may be called a pre-individualized society, a society which is so profoundly organic that its behaviour may be called psychologically communistic. There is no king in such primal cities but the priesthood is merely the brain, the ordering functionary which in the name of the God (who is owner of all and provider for all because he is the hypostasis of the community as a whole) administers the company of unlimited liability to which all belong, all contribute, and from which all are sustained. This form of society is immensely stable. That fact, however, is not wholly advantageous. If humanity is evolving psychically then not only must degenerative changes into egotism, specialization and unawareness be guarded against; progressive changes must be fostered. The task of a dynamic community is, by bridges built ever more purposively, to keep the two sides of man's nature in touch. As analysis grows it is even more important to have an analytical understanding of intuition than it is to have an intuitional understanding of analysis. To-day many physical scientists have the latter;[1] hence the advances in their understanding of the outer nature of things. Hardly anyone has begun to have the former hence the in-

---

1. See Henri Poincaré, *Science and Hypothesis, and Joseph Needham, The Sceptical Biologist.*

effectiveness of the good and even of the holy. At the beginning, beauty and use are blended and undifferentiated, as the bud contains both flower and fruit; in the end use and beauty have both to be distinct but not independent. They have each to reciprocate the other, use ever more efficaciously providing means, and beauty ever more clearly illuminating ends.

Up to the present the neglect of this reciprocity has constantly endangered civilization and humanity. Indeed, as we survey, on the large scale which is now possible for us, mankind's specific advance, it would seem as though the Quantum theory ruled as absolutely in social psychology as in physics. All changes are sudden mutations. There are immense periods when there is little change and, what in actual fact is more grave, whatever change there is becomes less and less significant of life. Indeed, there is nearly always that further aimless complicating of the social pattern which is, in society, a degenerative change analogous to accumulation of tissue and elaboration of form which marks physical evolutionary degeneracy. Then, when development has completely ceased, there follows an epoch of repression, succeeded inevitably by a cataclysmic uprush.

At present this uprush can best be traced not in Sumer, but in the twin source of our culture—Egypt. The mutation which transforms a stable, organic society into a radically new psychological form has first been detected in the Nile Valley grouping. Dr. Frankfurt believes that this change is symptomatized when the Egyptians suddenly passed from a cultural level at which there are no dated monuments (and perhaps we might add, no deliberate inventions) to another level where dating becomes of supreme importance, the centre of social activity. This fact, that dating, which till then had been incomprehensible, suddenly becomes obviously vital, leaves little doubt that such a change is wholly dependent on a new sense of time, and that, again, on a further precipitate condensation of consciousness. Individualized man is being crystallized out of organic society. A new, self-conscious creature is created, a creature which will require a society equally explicit in its regulations if it is to hold a being suddenly become explicit and peremptory in his demands. Nor is this being simply one who, because he feels himself precipitated out from his fellows (and sees the same separation taking place in his fellows), discovers that he is a stranger with his contemporaries. He discovers that he stands sundered also from past and future on an ever-shrinking moment of time. "He looks before and after, and sighs for what is not," realizing with the shock of a marooned man, waking from a dream of his home, that close behind him is an abyss of time, when he was not, and almost at his feet opens another abyss when he will not be.

Of this change from organic society, united by stable, social suggestion into groups of individuals, acutely aware of their separateness and therefore needing to be held together by greed and fear, by rewards and punishments, we have archaeological but not documentary evidence. We can see that there has been a sudden radical shift in man's interests, that he has in a moment become aware of a new world. We cannot know how he felt, how he endured the moral vertigo, as values, which he had so deeply assumed that he never defined them, vanished in air, and cravings which he had never imagined drove him to defy the social sanctions he had worshipped. We may employ William James's simile for religious conversion but reversing it, because at this historic moment the process James describes was reversed (and that is why to-day conversion takes place, and must, in an effort to restore a lost balance). We may say that suddenly man's nature, like a ship in a storm, "turned turtle" and all the individual animal physical forces which had been under became dominant and the psychological and social forces which had ruled were submerged.

The emergence of organized war, with slaughter and dismemberment of captives,[2] shows how violently culture was shaken and what a toll of life the psychological unpreparedness of humanity exacted of the generations which endured this transition. Man had made no preparation to balance this sudden condensation of consciousness, this shrinkage into individualism and loss of the intuitive, animal sense of kinship. He had devised no method whereby he might give himself a new, explicit realization of his union with his fellows, as explicit and as

2. See the Victory Panorama carved on Pharaoh Narmer's cosmetic-palette.

new as his new sense of his self-consciousness. Hence, as he emerges into clearer consciousness of the mechanical aspect of the outer world and of its manipulations, he sinks lower than the animal, through his loss of direct apprehension of union with his own kind. He sees no way of saving himself from falling down the slope of extinction through "intraspecial" struggle, through losing the sense which even animals preserve—that at least individuals of the same species are members one of another, and that war between them is war between cells of one body, that is, cancer.

Nevertheless, although this convulsive contraction of consciousness led immediately to great physical misery, to an acceleration of the divergence between man's powers and his ends, and a rapid advance toward complete individualization and specialization, civilization did not wholly disintegrate. Many possible developments were cut off. For if enlightenment was in the end to be attained by our race, it must henceforward be by much more intentional and agonizing effort. From now on intuition would be lacking; conflict would always be present; doubt and disheartenment constant companions. Such must be the future. In the immediate present the social order which had conquered was a brutalized order. The end, however, was not yet. The final issue was still undecided. If from now on there was not enough natural vision to deliver man into a new psychical life, neither was there yet a sufficient manufactured power to destroy him as a physical species. Mankind had taken the wrong turning, a turning which ultimately leads to death, but the quiet, degenerative return to animal level was debarred him. He must continue until by some daring revolutionary effort he can recover the true path of his evolution, or, by the continued increase of his means at the expense of his ends, he may generate sufficient uncontrolled power to destroy himself, not merely as a spiritual being, a creature still containing possibilities of evolving completely new faculties and functions, and able to live in a radically new order, but as an animal even of the most brutalized sort. It seems possible that man, with the attainment of civilization, with the discovery of time and self-consciousness, passed out of the animal consciousness irrevocably, so that whatever else befall him, gentle decadence, the return to complete unawareness, is denied him. Even at animal level there is much biological evidence that evolution seems to be irreversible. The only way out of any committed condition is onward. The doors by which entrance is made are built up behind the entrant.

We must, therefore, presume that although, as has been said, man devised no method which would give him explicit realization of his union with his fellows, he was able to devise what may be called moral moratorium methods. Reorganized religion backed the armed king, and so a State church, by the continuel employment of rituals, which had once given men direct sense of their common union, immobilized the natural reaction of anarchy against tyranny.

We have sufficient evidence of such a "concordat" from the immense place which religion never ceased to hold in Egyptian life. We have also sufficient evidence that that religion, from the close alliance between it and the armed State, becomes increasingly part of the coercive machinery of the State, ceasing to be a method of keeping the individual in touch with the common life shared with his fellows. Such a religion we should expect to prove not merely subservient to the armed power but also blindly traditional. It would not only lose the leadership of spiritual thought; it would actually sink to the point where it would lag behind the feelings of those who had no professional interest in understanding. The danger of clinging to forms would mean that at length the traditional would become first the inapposite and irrelevant, then the unseemly and even the indecent.

This is what actually befell the major official cult of Egypt, the worship of Amen. The religious revolution of Akhnaton was the consequence. Another sudden crystallization of consciousness, similar to that which began history by making man time-conscious, took place among the governing class in Egypt in the fifteenth century B.C. In this case, however, the event is later in time and so, not only have we documentary evidence of this mutation, but we can see that history does not repeat itself; that this later mutation produces a more intensified consciousness than does the earlier. In the earlier, man becomes interested in time; he feels for the first time his separateness from past and future, and from his present fellows. In the later mutation he becomes acutely aware of the absolute difference between himself and his deity—the wholly other *(ganz anders)* reigning in the sky, "to whom the heavens are not clean" and who therefore detests the fecund earth and the foul body.

Of this second revolution or "step-up" in the evolution of consciousness we have no clear contemporary comment which would throw light on the psychological factors involved. We can, however, from the actual moves made, assume that the traditional and established religion of Amen-Ra (an earth fertility cult on to which had been tacked the sun-worship) had failed to keep pace with the changes going on in men's minds as the tension between consciousness and unconsciousness heightened. The spontaneous conviction that there was a supreme individual Being in the sky seems to have drawn its emotional force from, and expressed the emotional need roused by a suddenly intensified separation of the two layers of the mind from each other, so that the outlooking mind, which now felt itself to be solitary (and that this was the only possible mode and type of consciousness) was isolated completely from the deeper layer of mind. This, as it actually runs the body and so is always aware of the body as its expression, controls that expression but, naturally, can feel no disgust for it, any more than the surface mind can feel disgust at thought in itself, but only at uncontrolled thought. We may also suggest that the traditional methods which once had served to provoke an expansion of consciousness, failed to work as consciousness became more focused, more discrete. We know that established religions have seldom, if ever, troubled to keep their psychological methods up to date; on the contrary, that they have shrunk from any such conscious effort. When they find that the traditional ritual no longer gives their worshippers the realization of being in contact with eternal life, they fall back on trickery to deceive the weak, and violence to coerce the stubborn. We know also that Akhnaton's reform went according to plan, that is to say, like all religious reforms it miscarried. Here again we can speculate that the very individualization of consciousness which caused this revolt from the traditional religion, making its fertility rites seem incompetent, superstitious and immoral, also made those who experienced such self-consciousness, incapable of putting anything else in place of the outworn religion, except an anthropomorphism (or zoomorphism). This anthropomorphism was nothing but a projection of that individualism which, while sterilizing the previous unself-conscious religion, was itself only able to put in the elder religion's place something more respectable, no doubt, but even more untrue.

Religious reformers are not, however, without use to religion, and the sudden changes in consciousness of which they are symptomatic, do in the end find appropriate expression. The symptom cannot cure the condition but it often heralds such a cure. Reformations generally fail because they are negative, protestant, and even when, as in the Akhnaton revolution, they put a complete new system in the place of the old, it is a system without roots. They do, however, stir the old religion, which they have cut down, to look to its roots so it may see whether it can reform itself, rendering out of its outworn and repulsive forms a technique which will give a psychological, contemporary equivalent of the older union with Life. This certainly seems to have been the course of mental evolution in Egypt. The stock did flower from below the graft. The Aten reform perished utterly; there was not, however, a return to the old religion, but only to its name. A new development sprang from the ancient tradition. From the fertility religion of Amen and Osiris, based on a belief in some creative depth in the springing corn roots and in the human body, a new and keener attention defined the new faith. There was a depth in man but it was not material or physical, it was mental and subconscious. The sharp division which Akhnaton (and as all reformers since) had tried to make between God and man, mind and body, was a mistake. The Mystery Religions were to spring from the development of the worship of Amen, Osiris and Isis—the Mystery Religions which were to carry man on to profound experiments whereby he should attempt as in Greece to make a system which would permit the most self-conscious intelligences to have direct awareness of their eternal life. This is not to sink the intensity of consciousness back into the quiet depths of an animal serenity experienced by the subconscious, but rather to raise the subconscious until its vast expanse of twilit awareness is all illuminated to the intensity of full self-consciousness. Moreover, the energy so to illumine and expand consciousness must be drawn from that fundamental force which is not bodily or sexual, but which has expressed itself in the past in physical form, partly, and partly in animal feeling, but which is still not wholly spent, and, if not wasted in the aimless repetition of past forms of expression, will, by reuniting the divided and weakened consciousness, give rise to a new quality of awareness, a radically new apprehension of a reality unperceived by the animal.

At this point in our attempt to give a brief psychological history of Egypt and the Near East, we face an additional difficulty. We have reached the time at which it is probable that the Mystery Religions took that form in which they were to represent, and framed that technique with which they were to influence, the higher religious life of the Græco-Roman world. Up to the rise of the Mystery Religions, religion is a public thing. In its evolution, therefore, although we have to speculate as to its actual spirit, we have its fossilized body from which we can form some estimate as to its functions and aims. Even after Akhnaton's revolt, the reformed religion of Amen, though the upper part of its spiritual spectrum is hard to detect, did have, we know, such an advanced front and, by the modifications in the explicit, exoteric and official religion we can estimate the energy, the direction and the progress of the esoteric development.

The rise of the Mystery Religions would, however, seem to mark a distinctive break. Indeed, we are probably helped in trying to trace this development if we see in Egyptian religious history (and maybe in all religious history of the same phase) three stages. The first is the period of conservatism. Religion is then a fossilized survival of the pre-chronological, pre-temporal culture and way of life. It makes no attempt to be a contemporary aid to time-conscious man in the solving of his present difficulties. He can take it or leave it, and all it provides (offers is too conscientious a word) is a crumbling wellshaft, down which he can clamber or tumble, to an earlier level of consciousness, there to find a pool of foul water. The second, post-Akhnatonic stage, is the period of reform. Organized religion does attempt to provide a private interpretation and probably an esoteric practice for those increasing numbers who require a method which will meet their contemporary phase of consciousness and the questions which this incipient individualism raises. This stage is one in which the doctrine of an individual future life becomes predominant. As each individual has done well or ill he will, as an eternally separate individual, prosper or suffer in a material next-world.[3] The third period is that when religion advances still further. Instead of being content to provide additional assistance, a higher way for those few who demand it, it is religion now which begins to set the pace. Anyone and everyone is called on to follow this, the higher life. Such an advance necessitates a break with the State Religion. The Two Ways of Life, a doctrine always symptomatic of religions in their second period, is discarded. The profitable connexion with the government is abandoned. Religion in this third period has renounced the attempt to make sporadic individuals less discontented with this present world and with the social order, through promising them a world to come. Its aim now is to help these spontaneous outcrops to develop into a new type of human being, the firstfruits of a mankind possessed of radically new insight and power.

This third period can of necessity be traced but very faintly. The connexion between public, official, exoteric religion and the esoteric religion is broken. The outward religion again fossilizes, content to have achieved that puritan other worldliness which makes its rituals respectable. Its alliance with the State is undisturbed, for with its convenient doctrine of the Two Ways of Life, individuals who might show an inclination to develop into the next stage of consciousness are to be persuaded to concentrate their attention rather on preserving their individuality in a next world. In brief, the third period of religious development in Egypt may with some degree of certainty be correlated with that wave of "personal religion" which Rhys Davids in his authoritative studies of Buddhism calls the greatest religious revolution in the history of mankind. We can now trace in the

3. The historical development whereby in Egypt the right to a future life was extended from Pharaoh, who first alone had it, to his great nobles and finally to all who could provide the funeral apparatus has been traced by Perry and others.

eighth and seventh centuries B.C. as it were a precipitate condensation of religious thought, spreading rapidly or (as Rhys Davids thought) appearing spontaneously along a line which extends from north China to south Italy. This "prophetic" movement breaks with the official, traditional religion. We shall, in the Hebrew record, be able to add a few details to our knowledge of this individualistic enlightenment and seership, as the Jewish canon was largely shaped by editors and authors who had a genuine if narrow sympathy with prophetism. We have, however, at best a scanty record, for a development which leaves no outward traces—in ritual and shrine—and in which, moreover, secrecy was imposed on the initiate, can hardly be a subject of historical study. Even if the proficients had desired to leave on record a full description of their procedure it would have been very difficult for them to have rendered in analytic terms the process whereby expansion was attained. Such a description is doubly difficult because, we must remind ourselves, the account would not be of a vestigial process but of an emergent faculty, and vestiges because they are retreats always must have a full vocabulary whereby they may be described, while rudiments because they are pioneer efforts have no adequate words in which their original experiences, functions and aims may be descriptively expressed. These men were attempting, and had to attempt, not a respectable way of returning to unity with life. They were endeavouring to take part consciously in a new expansion of consciousness, to advance to a condition of mind as superior to self-consciousness as that condition is superior to the undefined, restricted awareness of the animal.

Yet we have two guides which help us where to look and how to decipher what we may find. In the first place we know that development of self-consciousness was taking place. Man's dissatisfaction with traditional, official religion proves that. Secondly, we know that that development of self-consciousness, which caused man to abandon the old religion, would drive him to attempt to expand his consciousness by discovering a new religion. Further, there is evidence that the Mystery Religions which became famous in Greece and won the adhesion of nearly every eminent Hellene, were all derived from Egypt or from sources on which Egypt had drawn: a common and conveyable procedure must then have underlain most if not all the Mystery Religions, and what we discover about one, may well apply to the rest.

With this preliminary survey we can then proceed to put out our fragments of evidence along this outline. Though they are late, we have a series of references from Græco-Roman authors who underwent initiation in Egyptian Mystery Religion. Though forbidden by the invariable vow of secrecy to give an adequate description, yet many important hints are given. Apuleius, who in the second century A.D. went to Egypt to obtain initiation, records that during it he saw the sun shining at dead of night, which seems to indicate some intensity of light, an experience which seems to be the first reaction of the senses to an experience of enlargement of consciousness. Stobæus, a fifth-century writer, more closely notes the mental states passed through: "In initiation the mind is affected and agitated as in death. First uncertainties, labourings, wanderings, darkness. Then horror and trembling. This scene (or condition) passes and a divine light displays itself." This individualistic or private salvational method to give expansion to consciousness we can trace more fully in Greece. These authorities show how late an hieratic initiation could be obtained by those who sought it.

Beside such references, which can also be supplemented by comments such as Clement of Alexandria makes when, as a Christian bishop of the second century, he wishes to show that Christianity was the fulfilment and perfect development of the Mystery Religions, we have also evidence of the way in which exploratory religion developed in Egypt after the official priesthood themselves had ceased to research, and experiment in practical psychology had passed to unofficial bodies. Of what was the standard of that research, we have two fragments of evidence—one historical, the other anthropological. The historical evidence is from the Hermetic writings. There we see proof of a further development of religious experimentation. Up till then the highest stage of this exploration had been the private person seeking his individualistic salvation through an initiation given to him by an esoteric but still an official priest of the established religion. The transition from this to the succeeding stage we can actually place and date within a few generations, owing to a single classical reference. We know the locality and time where and when psycho-physical exploration was abandoned by the official priesthood, and Egyptian established religion (though, as we have seen, it was still prepared to cater for a few who needed what the past could give) relapsed into formality; thereafter all research being carried on by unofficial bodies. Strabo, the geographer, visiting Egypt in the last quarter of the first century B.C. says (Geography XI, 29, 806) that he saw at Heliopolis, the centre of Egyptian sacerdotal organization, the buildings then occupied by ordinary priests, for the specific trainings *(ascesis)* had been abandoned, but where the philosopher priests who had understood and undergone that training, had lived, and where they had taught Plato and Eudoxus when they visited Egypt.

The succeeding stage is when pioneering individuals, seeking for expanded consciousness, take two definite steps: the first is that they break away from the official temple religion—even an esoteric interpretation of tradition will not serve their need any longer: they have to be free to advance without any trammels of tradition. The second is complementary to the first: they cease to be solitary seekers and come together to form social patterns of avowed intentional living. The actual techniques used by the Hermetics are not given in those of their writings which have survived but we know from these three things: that they had techniques derived from India, that they generated their energy for the dilation of consciousness by sex sublimation—theirs was a celibate order—and that by living a collectivist life they formed a deliberate and avowed pattern of intentional living. This third factor is of increasing importance in the attempts made by mankind to obey the urge within themselves to a further evolution and to give that evolution untrammelled and adequate expression. As Buddha had said: "Right Livelihood" is one of the essential steps in the Eightfold Way, for if we fail to find a way of living which will aid our every thought, word and deed, we are carrying an almost impossible handicap in any attempt to dilate consciousness.

Yet although the Hermetics diverged completely from the official, hieratic, contemporary life of Egypt, they did not, it would seem, make a complete break with their religious heredity. Their books we know were said to number forty-two (perhaps one for each Nome of Egypt) and to contain all the Temple rituals. They claimed as their founder Thrice Great Hermes, or Thoth the apotheosized man-God of Egyptian Wisdom. They were then, a development, not a breakaway, as also were their spiritual successors, the Essenes. The attempt to advance further the Hermetic group experiment is, we may say, with some certainty, the Essene community. As this, however, was attempted in Palestine, it had better be considered in the next chapter which deals with the Hebrew efforts at the furtherance of conscious evolution.

Another contemporary development which is parallel to the Hermetics is that of the Therapeutæ. Unfortunately our sole authority on this organization is Philo, who as a Platonizing Jew in his *De Vita Contemplativa,* naturally wishes to describe a community which believes that petitionary prayer, intellectual study of a verbally inspired text, and asceticism are all that is necessary to attain enlightenment. Yet Philo owns that he does not know when this order started or what precisely is the meaning of its name, Therapeutæ. It may signify simply the Servants of God or it may connote the physicians, the new psychotherapeutists who could cure the conflict in the individualized consciousness by releasing the thwarted spirit upon the upward path of its evolutionary deliverance.

Jewish thought had as we know attained unprecedented freedom in Egypt (only, at Elephanta in Egypt, did Jews ever build themselves another Temple): the Ptolemies were philosemite and in 160 B.C. a Jewish forerunner of Philo, Aristobulus, began the attempt to reconcile Judaism with Greek thought. Yet that a social heredity, as alien from celibate monasticism as is the Hebraic, should have given rise to a cœnobic order, without any Egyptian or other religious influence, seems unlikely. We know that the Hermetics themselves drew upon sources outside Egypt itself. Flinders Petrie in his *Doctrine of the Future Life in Egypt Prior to the Introduction of Christianity* has shown that in the Hermetic writings there are open references to the Indian masters to whom the Hermetics themselves looked as far advanced religious explorers and experimentalists.[4] It therefore seems probable that Philo makes his account of the Therapeutæ as much as possible a piece of Jewish propaganda to show that Judaism had not stopped still but that it, too, could not only contribute but that its contribution was supremely important to continuing the growth of religion. He, therefore, shows his Contemplatives praying to an external God at sunrise and sunset and spending their entire day in study of the Jewish Law (the Torah). They meet on the Sabbath to hear a discourse and to sing hymns. All this is quite in keeping with formalized unexploratory religion. There is no psychology here except the most vestigial, unconscious sort. Yet this rationalized, synagogized religious practice is supplemented by certain psychological methods which suggest that the system which Philo described was more than a Fundamentalist college. The ascetic discipline was so severe that the members must have followed it for some reason other than the pleasure of rational restraint. They did not believe that their God was pleased by their suffering. Neither of their masters, Moses or Plato, had taught mortification or even definite asceticism. Yet these quiet students, each living in his own small house (they are described as living near one another in Alexandria) were evidently fascinated by experimental abstinence, none eating or attending to any physical function save at night, some eating only every other day, and a few eating once a week. The food was coarse bread with salt, the drink water. We know that fasting, however inaccurate and unsound a method, is the commonest whereby man provokes a disturbance of ordinary consciousness. The Hebrew canon witnessed to its effects: Saul "ate no bread all the day nor all the night," before the vision of Samuel given him by the Endor witch. The fast of forty days was believed in Arabia to give power over the Djinns who became visible to the faster. Many Yogins fast until the Gods are seen face to face. The medieval, proscribed mystics, the Bogomils, fasted until they saw the Trinity before them and an early tradition supported their practice by telling that James the

4. The date of the Hermetic writings can also be estimated as the word "Satrap" is used, a term impossible before the Persian conquest of Egypt under Cambyses and unlikely after the fall of the Persian Empire in 330 B.C.

Less, to obtain sight of the risen Lord, had to vow a fast which, after it had been maintained rigorously, he saw; while the almost canonical book, *The Shepherd of Hermas,* advises fasting which will be rewarded by a "Vision of the Church."

That such a method, though violent, imprecise, Fakiristic, is not, however, wholly unscientific, wholly without experimental control, we know from its method of employment among some Amerindians. Fasting to obtain enlarged consciousness is regularly employed but the visions are not uncritically accepted. They all have to be written down and then the Elders of the tribe wait with these records to see whether references to the future will be confirmed. Only if they are, is the visionary accepted as a Seer.

We may believe, therefore, that the Therapeutæ were rudimentary experimentalists for not only did they fast, but each in his small house kept a room reserved only for his Sacred Studies. They were attempting extreme concentration and to aid it, they employed the powerful assistances of regular hours and a reserved place. Philo undoubtedly thought that by these studies (which, as they were loyal Jews and Hellenic intellectuals, could only be allegorical interpretations of the Torah made to give intellectual conviction of a Supreme Deity) the Therapeutæ did attain a higher enlightenment than any religions that he knew. Indeed, he maintains that in comparison with their pure Contemplation, their constant and sole irradiation in the Light, the Essenes were only secondary, because they were "practical," being concerned also with their physical subsistence. How the Therapeutæ were provided for, that is not told. Members joining became, as later did the monks, legally "dead," losing all property, but as the Therapeutæ seem to have been "retired" men it is possible that they, as the monks after them, were able to endow their order with some of the funds they had made in earlier life and now personally renounced. In any case, whether by their own gifts or those of the charitable, the Therapeutæ had to be kept. Philo, therefore, has no right to despise the Essenes on the count that they kept themselves; while as to actual psychological achievement we shall see when we come to them, that the Essenes seem to have been more advanced than the Therapeutæ—more advanced in social organization, in freedom from tradition and in techniques.

In short, the Therapeutæ, if we are to trust Philo wholly, seem to have been an organization which attempted to free religion from anthropomorphism, to advance psychology and to widen consciousness. As a matter of fact it reverted to actual methods which are only to be placed as advanced Fakiristic: methods which lie, evolutionarily, between the simple dervish Fakir such as we shall see in the following chapter was Elijah, who blindly exposes himself to every natural hardship, and the rudimentary Yogic methods of precise techniques for dilation of consciousness such as are recorded in the Hebrew canon as being used immediately after Elijah by his successor Elisha. This traditionalist failure of the Therapeutæ had grave consequences. For from their superstitious reverence for the Hebrew canon and their use of documents quite unsuited as mantras (as texts by which the mind may rise from meditation to contemplation, to true dilation of consciousness), arose, as Eusebius recognizes, Christian Egyptian monasticism. This, we shall see, increasingly lost all courage of experimentalism (at least the Therapeutæ wrote a few new psalms for themselves), degenerated further into masochistic Fakirism and so brought the history of man's psychic evolution in Egypt to an end.

This chapter can be concluded with an examination of anthropological evidence which throws light on Egyptian methods. Studies of the magic and ritual of Africans has (see Seligman) in the last few years established with some certainty that all the systems for the disturbance of consciousness, practised by the African Negro, are derived from those of ancient Egypt.

What such systems can effect we are beginning to understand from those anthropologists who are prepared to learn from the native, and not merely to explain away his methods.

Loram, the commissioner of the Trans Kei Bantu reservation, in 1925 had discovered that the witch doctors of that part of South Africa are possessed of a technique for developing extra-sensory perception by putting themselves into trance without the use of any drug.[5] Further, he found, they are able to select boys who have this faculty but lightly buried and to show them how to bring it under the command of their conscious will, and it is of some possible interest to this inquiry, where no clue must be neglected, that the witch doctors (of course unacquainted with the European medieval tradition about the same thing) select always if they can find such, the seventh son of the seventh son. The fact that small families become common when psychic faculties become rare, may then not be a purely fortuitous association. Genetic research among humans has already shown that there is a real basis of fact in the tradition that the children of older parents tend to have higher intellectual capacities.

Loram studied these practices only from without. Gorer *(Africa Dances)* was actually initiated. On finding that his faculty for extrasensory perception had functioned sporadically, the priest-king, Aho, in French West Africa permitted him to be initiated into two totems and to see the methods used by the witch doctors for obtaining information by hyper-normal means. The powers at the disposal of these men divide roughly into three classes: First, the totemistic method or direction. This technique gives control over animals and even, it would seem, over some activities which we call inorganic (though lately it has been said in the foremost scientific journal, *Nature,* that the distinction of organic and inorganic is becoming increasingly arbitrary). The second class may be called the physiological methods for giving hyper-normal control of the body. In this class we have anæsthesia, hyper-normal resistance of the body to injury and suspended animation. The third class may be called the specifically psychological methods whereby—as Loram also witnessed—the consciousness is detached from the body and collects information beyond the range of the senses. From this anthropological evidence we can form some general idea as to the outward manifestations of Egyptian psychology.

It is important to note that these distinctions of psychic faculties are recognized in Yogic practice: one Yoga having to do with the control of animals: the control of solids being confined to lower caste Yogic practice, while that of matter in other states, such as fire, being reserved for the Yoga of Brahmins.

---

5. cf. Chap. V, on Greek methods at Delphi.

We are also making a legitimate estimate if we presume that the Egyptian techniques and results were more accurate and extensive than these methods now practised secretively in Africa. For it is not possible to avoid the conclusion (a conclusion which Gorer himself suggests) that the methods for the dilation of consciousness employed by Negroes to-day are decadent. In support of this supposition it is enough to recall the height of social order and the loftiness of moral responsibility which existed in Egypt when these techniques were in use, and to remember how Loram and every other informed inquirer have remarked that wherever to-day extra-sensory perceptual techniques exist, the techniques are vague and the motives generally gone black. Loram was led to his investigations because the witch doctors were using their powers for tariff-rated murder—one gold sovereign was all that was charged for killing a man. Gorer remarks how incessantly blood was used to aid the techniques. Both authorities agree that the whole knowledge is declining and that as soon as a native becomes acquainted with our analytic culture, he despises and abandons all these psychological methods.[6]

Such then, is what we may learn of the development in Egypt of conscious psychological techniques for the expansion of consciousness. We may say that there is definite if scanty evidence of a deliberate attempt to find methods whereby the latent energy in man might be given an expression which would not merely be a catharsis, a deliverance from balked disposition, but lead to new powers and new vision.

6. It may be wondered, however, whether as the native finds analytic culture unsatisfying and sees the civilization based upon it losing prestige and crumbling, he will not return to the older methods. We have seen that such a springing from below the graft marked the post-Akhnaton reform in Egypt itself and, in Chapter VIII, on the Greek development, we shall see a similar return.

CHAPTER VII

THE HISTORICAL EVIDENCE FOR THE EVOLUTION OF CONSCIOUSNESS

II—Hebrew

OF THE three influences which have blended in our past, Egyptian psychological development yields the earliest clues. We can find in the Nile Valley traces of two great preliminary steps in the evolution of consciousness. We have discovered an imprint of the crisis when man became conscious of Time and so realized his utter individuality. We have documentary evidence of that later mental revolution (Akhnaton's) when the break which began with the sense of separateness from his fellow man, was completed by a sense of separation from his God, from the creative power and its presence within him.

These records are of great importance to the understanding of that phase of evolution when psychic development begins to take priority over the psycho-physical, as the main line of evolution. Yet in these changes we have no indication that man was conscious of what was befalling him, that he understood his fate and how to co-operate with it. His mental evolution was still as subconscious as had been his physical evolution. Of the time when he began to understand that changes which he had assumed were changes in his circumstances were really changes in himself, we have scantier evidence, although the date is later. When man begins to work out a technique whereby he might help release and express the subconscious forces which were pressing him, he does so as an individual who must work secretly. Whether within the official religion, or as later with the Hermetics, outside of it, man explores, as it were, underground and as a lonely pioneer, leaving the main masses behind. No heaped hills of pyramid and temple indicate at this stage where the mole of the human mind is burrowing its way.

It is, therefore, of the highest value to our inquiry that, as the Egyptian path of exploration goes wholly underground, another people, and one unusually interested in recording its religious development, should take up the story. It is further of peculiar importance that the Hebrew development should derive directly from the Sumerian and Egyptian and that it should branch off as an independent exploration at the very time that the growing front of Egyptian religion was becoming secretive. Because of this Semitic interest in Time, the Hebrew record has the additional value that it can be dated with considerable accuracy. We see by this that it fits between the Egyptian and the Greek developments and we can watch, during the phase when it made its distinctive contribution to human psychic evolution, the explorers of Israel, generation by generation, realizing more consciously how they might cooperate with the force seeking expression in them.

The specific Hebrew development begins under Moses, though every historian now recognizes how much already must have existed as a religious tradition among the people he led, and, further, how much of what afterwards became canonized as his teaching also contains later contributions. It is probably true that this leader was trained in Egypt, although it is strange how little that is specifically Nilotic appears in the ritual of the Tabernacle. Nevertheless, Moses stood in the minds of his people as pre-eminently the monolatrist, and it was this doctrine which he is most likely to have acquired in Egypt when the monolatry of Akhnaton was still the highest stage of spiritual thought and before the re-established Amen religion began to provide in its system some psychological technique to meet the needs of those who desired a development and expansion of their consciousness. Something of Moses's psychological exploration we can learn from Exodus, in spite of the fact that the book cannot be, as it stands, a contemporary account and has certainly been liberally edited by men whose religion had become extremely "projected"; men who were increasingly convinced of the separateness of their Deity from themselves. Moses we know was married. He could not, therefore, have used complete sex sublimation as a technique for enlarging consciousness. We also know that he left the priesthood to his brother Aaron. Granted, then, that the three original pre-scientific methods of dilation are: the Shamanistic, by violent movement often aided by drugs producing dissociation; the Mortificatory, or the Fakiristic, by pain; and the ascetic, or Raj Yogic, by sex sublimatory suppression, in which of these categories should we place the man who put so lasting a stamp on the religion of the Hebrews?

Those who have read Dr. S. H. Hooke's important essay on "Initiatory Rituals" (in the volume *Myth and Ritual)* may be inclined to assume that Moses belonged to the Shaman medium type. From the account given of his life there does not, however, seem to be sufficient evidence to show how he strove to obtain dilation of consciousness. Indeed, it would seem highly probable that as a monolatrist he did not have any deliberate technique, being confined to petitionary prayer. The ritual of the Tabernacle, as given in Exodus, is generally considered to be considerably later than the Exodus itself—in fact, an imaginative sacerdotal construction made by an author who knew the late Temple ritual. Such a ritual, whether known to Moses or not, cannot be thought of as a conscious technique for transcending the limen.

Another psychological problem is raised by the "Mosaic" code and must be considered here. Whatever the code's date, it confirms what is certainly an old Semitic tabu, the tabu against any private or unofficial attempts to explore and expand consciousness, the commandment against witch-craft and all soothsaying, a law the breach of which was death. Such a law with such a penalty seems to spring from the anthropomorphic stage of religion (a religious position which as we have seen was quite possible for Moses) when all experimentation in psychological things becomes blasphemous trespass into the presence of a personal, arbitrary God. Certainly this tabu was being stringently enforced by the rise of the kingship, for Saul is said to have destroyed the witches so thoroughly that when he found himself in need of a medium he had difficulty in finding one.[1] We see, then, the narrow limits in which at the beginning of the first millennium B.C., Semitic tradition permitted any psychological contact, still less explo-

1. Samuel, xxviii, 3, 9

ration or research—limits which would be steadily contracted as earlier animistic intuitions about religion were starved and crushed out by anthropomorphic thoughts and theories.

Nevertheless, a dying gleam of native insight continued up from the fourteenth century to the eighth B.C. Such seership is essential and has to be accepted. For Ritual Sacrifice does not bring outward luck or inward wisdom, strength or peace. Without the true visionary, the man or woman who has direct contact with the larger consciousness (from which all individuals spring, in which they are all still rooted, but from which their acute individualism cuts them off from conscious contact) human society cannot consist.

The first reference to seership unassociated with military power and possibly associated with a definite technique of "dissociation," is in the Book of Judges. In this account, which is generally considered to be a Bronze Age document very little edited, we come across the story of Deborah and Barak, itself probably a twelfth-century story. Deborah is a prophetess—a woman who somehow convinces the disordered tribes of Israel that she had a mediumistic insight on which they had better depend. She is married, so her method cannot be "yogic" (sex repressive, for it is extremely improbable that she would have known of any of the Laya Yoga methods of using sex for consciousness dilation) (Chapter X), but her husband (Lapidoth) has no authority.

She herself gives judgment by her own authority of insight and when she needs to launch a military uprising against the Canaanitish King Jabin, she does not employ her husband—she is no priestess queen-consort. She summons the fighter Barak from some distance away. He will not advance without her magic to guard him, so she attends him to the hosting place and tells him when he is to launch the successful attack.

Such detailed accounts do not, however, give us any description of the prophetic technique. Nevertheless, if we look through these later historical books of the Hebrew canon, which is in this part arranged in a chronological sequence, we can watch a distinct development of the attempt to attain insight. In the social chaos which preceded the establishment of the centralized authority at Jerusalem, the author of the Book of Judges concludes that sporadic insights, such as Deborah's judgeship illustrates, are not enough. Society cannot hold together on such scanty sanction, still less find its way forward to a higher degree of consciousness, with nothing more than these random and faint sparks of illumination. The chronicler owns that the epoch of the Judgeship closes in moral and social chaos and spiritual bankruptcy. And his diagnosis appears to us now to be sound and inescapable. Society was perishing, man degenerating into anarchy because no one had, or knew how to obtain, convincing insight into an order of things above animal level, an order which would sanction moral conduct. Life was hell, "Every man did what was right in his own eyes" because "there was no open vision." It is always true of human society, whether in the Neolithic or in the immediate present, that without manifest apprehension of an order more consistent, more moral than that perceptible by the average sensual man, a lawful community is impossible. Such an apprehension may be intuitive in pre-individualized societies. The more self-conscious the constituents of a society become, the more definite and more constant must be the apprehension of its seers. The state which is based on physical sanction and power, and not on spiritual—whether it call itself a democracy, an autocracy or a theocracy—is never at peace but at latent war, horizontal or vertical. Provided the society remains economically unprogressive, with its knowledge of physics as arrested as its knowledge of psychology, the cake of a dried and rigid tradition, "the crust of custom," may preserve it by paralysis from anarchy. Sooner or later, however, spontaneous specialization and idle curiosity, symptoms of intensifying consciousness, will begin to work and the social order must crack.

The centralized authority which crystallizes about Jerusalem appears to have been formed of three factors. The most evident but least important was the kingship. The Book of Samuel shows the emergence of the armed autocrat All that the prophet Samuel had foretold about this form of authority came rapidly true. He was naturally much more effective in crushing his subjects than in subduing their enemies. Yet it is clear that he would not have been able to have had himself appointed as king had the prophet-judge been able to keep in constant touch with the larger vision which sanctioned moral conduct. Whether through slackness and indulgence on the part of the prophet-judges themselves (as the story of Eli would indicate, this was at least part cause of the decay of the spiritual authority), or through the general evolving intensification of self-consciousness, causing earlier psychological techniques to become ineffective, it is clear that the spiritual authority had become impotent. Had there been "open vision," peace and order would not have had to make their suicide pact with militarism and tyranny. However, the social squalor which the kingship only accentuated, brought about a reform, at least among the spiritual authorities. A split seems to have taken place, dividing them into two classes. On the one side was, probably, the majority. These became increasingly perfunctory and conservative and therefore had no choice but to throw in their lot with the kingship and to depend on the autocrat's armed forces for the protective sanction which they themselves could no longer exercise. Hence they sank to a level where they had to approve whatever policy he pursued and method he employed, for his arms were their protection and their spiritual pretensions mere pretence.

On the other side was the socially secessionist number which formed the schools of the prophets and Nazarenes—a type which seems to bear a similarity with the Fakir. This aspect of the spiritual class, relieved of all dependence on the State, evolved rapidly. Elijah is a Fakir; his successor, Elisha, a Yogin. This difference may be more precisely defined if we recall that the Fakir undergoes all and every austerity that wild Nature can inflict, his only science being a watchfulness to attempt to use the intense experience of such hardships as a means to dilate consciousness. The Yogin marks a great step away from this "field naturalist" attitude, toward that of a laboratory scientist. For Elisha knows how to produce his results and so (he lives in a house with a servant) undergoes no unnecessary hardships. We have, fortunately, a description of his technique, and it is closely in accord with that employed to-day by those who wish to bring about what psycho-physiologists call dissociation but which is actually extra-association. In order to fall into trance the prophet has music played to him, so as to soothe the restless listening of the ear, a sense which is more difficult to control than the eye's.

We have, then, clear record, certainly as early as the eighth century B.C., showing that some men had learned not merely how to attain extra-sensory perception but how to make the method almost a science. They no longer exposed themselves to indiscriminate strains, hoping that one of these might start a stave of the cask in which they were cooped and so, through the crack, they might catch a glimpse of the larger universe. They direct with precision and accuracy their attention on a certain focus, and their physical methods are confined to removing such interferences as that focus may otherwise suffer, from the common senses of physical apprehension.

Here, however, once again our evidence fails us, or it may be that the facts themselves do not exist. We know that there were schools of prophets, "dervishes," who had actual settlements (training places, we may assume) at the time of Elisha. The roaming bands of a generation earlier are now under some sort of organization. Perhaps it is under his orders that they build these centres, for one of the thaumaturgic stories, of his making an axe-head float, is told of the erecting of one of these settlements. But by another generation this development seems to have collapsed and the whole idea of the prophet to have undergone a radical change. The difference between Elijah and Elisha is one of development. They aim at the same thing: contact with, and emergence into, an order of reality which, being above the order which ordinary animal man may perceive, gives the prophet power over the lawless as well as over the morally minded. When Elisha wishes to reassure his servant he has the young man's eyes "opened," who then sees the surrounding countryside aswarm with burning figures. Elisha also acts as a telepathic secret-service agent for the king of Israel.

We are not discussing the authenticity of such psychic exploits. The point to observe is the complete break between this frame of mind and that which immediately succeeded it. The difference between Elisha and Amos (the dates of the two are scarcely a generation apart) is a difference of radical departure. Put crudely, Amos cannot work miracles and so can only appeal to conscience. Beyond that he can only threaten an oncoming, undated Day of the Lord which will be a day of reckoning and vengeance. He can stimulate what moral good there may be still present; he cannot alter the will set upon evil, nor even convince the wavering that behind the world of common sense there is a greater reality which sanctions moral conduct.

The later prophets, who descend from Amos and Hosea, and culminate in Jeremiah, claim, it is true, a supernatural sanction. They are not pure moralists, appealing to nothing but a mutual self-interest of irreducible individuals. They assert that they have had a revelation, their messages always beginning with the phrase, "The word of the Lord came unto me." Such a claim is, however, in reality very little sanction. It may stir those who wish to believe. It cannot hold any who are determined to go in another direction, or even inspire those who are (either as rulers or subjects) confronting a situation where physical force is holding the field. The king is as fully aware as the slave that only a miracle, only real power to touch a reality above physical fact, can save him from bloodshed. How is the law-giver to avoid becoming the arbitrary tyrant so long as right is only manifest—can only become actual, when violence is used? Place on the throne the prophet-moralist himself, the man who has no spiritual power but to stir the consciences of his congregations, and their consciences, faced with the fluctuations of their own natures and the obduracy of outer Nature, will relapse, his emotional appeal will fail, and he will have to use violence (perhaps more indeed than the tyrant, because the idealist's demand of his subjects will be higher) to impose his righteousness. This certainly has been the fate of all theocracies. In short, the later prophets, claiming as their sole sanction and proof of their messages' truth, "I have had a revelation," are on pathetically weak ground. As the wise, gentle but tired Selden was to remark as the last age of private revelatory prophesying reached its climax: "You may claim to have had a private revelation, but I must also have another, to know that you have indeed had yours." This is a stalemate.

It is just this difficulty which the post-Elishac prophets failed to face. They show no hint of understanding this problem: that the prophet must give his listeners some means, some psychological technique for sharing his insight. Instead, they attempt to establish their case, the truth of their message, by *post hoc propter hoc* arguments, by saying that all the political misfortunes which came upon the State were due to their messages being disregarded. This argument required much editing of the facts. Capable kings, of whom they disapproved (such as Jeroboam the Second), were successful. Even Ahab is not struck by immediate disaster. Good kings, such as Josiah, suffer defeat. These facts gradually force themselves even on the prophets. Their teaching becomes increasingly individualistic and other-worldly. They cannot make the deeper reality disclose itself, they realize. The only thing for man is, then, to await God's time, the Day of the Lord, that moment, when, his provocation being sufficient, he will intervene with apocalyptic transformation of the natural order. This other-worldly apocalypticism is an inevitable immobilizing conclusion, once Deity has been projected, once man has broken up reality into three irreducible units, Man, God and Nature. Once that is done then there is no escaping the final conviction that only the violent interference of an external deity can put a fallen world back into original perfection. If, however, this anthropomorphic projection has not been allowed to take place, if man has been able to keep before himself the realization that he and all objectiveness are finally one and that the problem of psychology must always come before the problem of physics, then he can see that the true issue is always whether he can alter the focus of consciousness and that if this can be done, a reality other than the animal's becomes visible.

With our psychological insight, we are then compelled to ask whether the old explanation of Jewish religious evolution is accurate? Is it merely the transition from magic to morals that makes Jeremiah a martyr, while Elisha remains invulnerable? Is it not possible that as anthropomorphism grew, though it gained in ethical loftiness, it lost in psychological power and even interest? Must we not inquire whether religious mankind, because they were increasingly certain that there was an external judge of all the earth, who would do right in spite of humanity's ignorance and ill-will, neglected to develop those rudimentary powers which would have given them direct apprehension of the universe's deeper order? It is impossible for us to say whether this arrest in development was inevitable. One of the greatest of historians, Lord Acton, used to warn his students against the easy assumption that what happened alone could have happened. Nevertheless, if, as is the hypothesis of this essay, evolution is now being carried on through human history, at immense acceleration and through the expansion of consciousness, then we might probably expect that though there would be progress, it would be confined to the very few, would break out at unexpected times and from unsuspected sources.

Such seems to have been the history of man's anxious but dimly understood endeavour to enlarge his consciousness. As in Egypt, so now in Palestine, as that endeavour becomes more deliberate and intentional, we find increasing difficulty in tracing its path. The division between the priest and the prophet is fairly evident in the Hebrew chronological books and in later references in the records of the prophecies themselves. Of the later but more important division, when the prophets themselves split into two parties, we can do little more than surmise. We have evidence of the prophetic party which relapsed. We have clear proof, for instance, from the final book of the Old Testament, the book of the prophet Malachi, that the ethical prophets, claiming a revelation (but no powers to train others into a deeper apprehension of reality), have gone back and made alliance with the priesthood. All that the revelation can now promise is that if the people will pay the dues to keep the Temple sacrifices going, God, pleased with their payments, will repay them with worldly success.

We have far scantier evidence as to what happened to the other party, which could not accept such retreat. As in Egypt, however, we know that there was a clean break away from the national life and the official religion. The Essenes parallel the development of the Hermetics. In both cases the seekers of that time concluded that the only way of continuing research into psychology, the only way to make deliberate attempt to expand consciousness, is to abandon the old life altogether and to attempt to build up a way of living in which the expansion of consciousness, the deliberate, intentional, experimental attempt to forward psychological evolution becomes the one purpose and aim binding together the social group. The Hebrew development seems to have missed out the intermediate stage when the advanced individual may have the popular religion interpreted privately for him by the priesthood—a stage lying between anthropomorphic puritanism (Akhnaton's reform in Egypt: Josiah and Hezekiah's reform in Judea) and the new intentional community, the frame and direction of which is psychological development (the Hermetic and perhaps the Therapeutic in Egypt and the Essene in Palestine). There is nothing in Hebrew writings that gives any evidence of the Jerusalem Temple religion developing esoteric techniques which might—as did the re-established and reformed religion of Amen in Egypt—supply the advanced individual with an initiation and training in the development of consciousness.

Of the next stage, of communities broken away from the national life and the religious tradition, in order to follow such development, the Essenes give us perhaps better evidence than the Hermetics. In the latter case we have some of their writings but lack any adequate account of their lives. In the former, as we shall see, we have no extant writings but a fair contemporary account of the social organism—the Essene society—itself.

This order does not seem to have precipitated at once into a complete organization, with an organization's three specific functions—a psychiatry, an economy and a policy. If we compare and attempt to harmonize the three historical accounts which alone approach any fullness—the accounts of Philo, Josephus and Pliny—it seems clear that we are presented with a development. Josephus says they "had no single city of their own but dwelt in every city." They are, at this stage, apparently scattered, like some masonic order, over a wide area and only bound by common vows and certain private psychophysical exercises. It may perhaps be assumed that at this date they are deriving from the "seeker state"; that period of inquiry, individual purgation and the seeking out of like-minded spiritual explorers and pioneers, which marks the start of every spiritual reform and mutation. We know that at that time the Chasidim had arisen, those associations of the "Holy" who were determined to see whether a life more scrupulous, pious and purged would permit the devout Jew to understand why God had not "restored the Kingdom of Righteousness" when the Maccabean revolt had succeeded and the Temple worship had been restored. The Hasmonean house had sunk to the level of any oriental despotism and there was intense discouragement and drastic searching of heart among those devoted men who had hoped (as many such have hoped mistakenly since) that in a drilled and disciplined nationalistic revival a practical theocracy might arise. The first reference to the Essenes is in the reign of Jonathan the Maccabee. It is therefore chronologically possible to see them as a development of the Chasidim, as their name (as does the Chasidim) probably means the Holy or the Pious. We know that the Pharisees were the decadent branch of the Chasidim—men who sank to being formalistic pietists whose real interest was a ruthless capitalism. In the Essenes it seems reasonable to see the developing branch from the common Chasidic stock.

To develop, and not to relapse as did the Pharisees, they had, however, to struggle hard. Significantly, they had to leave the cities which proved fatal to the Pharisees' *moral.* Therefore it is natural to find that Philo reports that he knew of them as men "who prefer villages" and follow agriculture rather than trade. By that time Philo estimates they numbered a scattered order of some four thousand men. That they were as well known as Quakers may be assumed from the fact that one of the gates of Jerusalem was called after them. This stage, however, could not last, and it would seem that it is the succeeding third, and final stage given by Pliny where he describes a definite colony planted in the wilderness by the Dead Sea.

It is to this completed organization that we can alone apply that entire cycle of rules, training, livelihood and ranks which our three authorities give. Josephus actually lived among them in youth for a short while *(De Vita Sua 2),* but clearly for too little a time to have undergone even a preliminary initiation. He reports, however, with Philo, the structure of the society and it is clear that here is an entire social organism, possessing a psychiatry which successfully trained the individual to a level of control unknown in the West, an economy, so that there was a complete and planned corporate livelihood whereby a society could live up to its psychology; and a policy, a consistent attitude toward the world, a complete and working answer to the question, "What is the meaning of Life?"

As we should expect, in such an organization there is a long and careful novitiate—it lasted three years. The life was to be completely purposive and intentional, hence the postulant had to accept the postulate that "Pleasure is Evil," amusement distracts from purpose. The vows were exacting. Complete celibacy was imposed; women were considered too dominated by desire and distraction to be able to take part in such an enterprise. Complete collectivism was also essential. Each member gave his entire goods to the community. Slavery was emphatically condemned. Hence the community had to be self-sustaining. The simplicity was austere. Though cleanliness was very high, garments and shoes were worn until they fell to pieces. The meals were as austere. There seem to have been only two—one at noon and another at sunset, both consisting of bread and one dish. Time was as severely rationed as food and clothes. They prayed to the sun at its rising and spoke no other words until it had risen when, after this prime service, the worshippers went to their tasks until eleven o'clock. At that hour they came in, washed and changed into linen, after which the common meal was taken. They did not impose silence, but because their minds were trained, they spoke only when necessary and with reflection, so that visitors who were admitted to such meals found the atmosphere awesomely quiet. The vows also included a strict promise—they despised oaths—"to keep secret the books of the society and the names of the Angels."

We may ask what were these books and names? Certainly some of the books, but almost as certainly not all, were the Torah, the Psalms, and perhaps the Prophets, in fact the Jewish Canon much as we know it. For not only did they read the Scriptures: they did not conceal the fact that they employed practices quite other than those of the Jews. They worshipped the sun, were strict vegetarians, and emphatically forbade animal sacrifice and would not even anoint themselves with oil, considering it unclean. Their cleanliness was vital to them. Careful washing, changing their garments before meals, and the wearing of white clothing seems to show a Yogic standard of hygiene. The books must therefore have contained instructions beyond, and indeed different from the Hebrew code of life as given in Deuteronomy. What may be meant by the "Secret names of the Angels" may be discussed when trying to estimate their psychological achievement.

Before dealing with that final issue it is necessary to point out that this community was a true and full social organism, not a mere common stock of goods and hands. It was not a communism or indeed a simple collectivism, rather it was a living body politic. The system consisted of four clearly-graded ranks or castes—with this vital proviso that it was taken for granted that, during his life and development in the community, the individual member could and should strive to rise through the order to its headship. Nor was this system one imposed. All elected those who were to rise, even to the highest offices of judges, priests and prophets. It was because you were able obviously, in the opinion of those who lived intimately with you, to sustain higher responsibility, because you had achieved higher controls and understandings, that you were called upon as a friend and for fuller helpfulness to all, "to go up higher." This view of rank is clearly revealed in the sensitive injunction to those in control as judges and priests that they shall take care not to outshine their inferiors on any occasion where they meet socially and as temporary equals. This fourfold ranking is naturally of great interest and highly suggestive. For the Laws of Manu, the basic Indian code, has four castes. The likeness becomes even more striking when we read that here too (but without the harshness which hereditary caste gave to it) was the principle of untouchableness. Indeed, it was among the Essenes, carried beyond the Indian custom. The three upper ranks could each be "polluted" by touching or being touched by any rank lower than the one at which they had arrived, and probably the lowest, fourth rank, was debarred from contact with the uninitiated. In passing, it may be noted, such specific regulations would have been impossible to the society while living scattered throughout the land, for pollution must be immediately purified by washing.

This complete but flexible caste system, together with strange details of worship so alien to the later Jews, such as adoration of the Sun, have always suggested to scholars that the Essene system was derived from some extra-Semitic source. It used to be thought that this could only be Persian. With our present knowledge of Persia's, and the whole Near East's, indebtedness to India, there can be little doubt that in the Essene community we have a fairly complete extension into the Levant of Brahmanic teaching and techniques.

That word techniques brings us back to the question of actual conscious psychological procedure. Were the Essenes, at their height, a community of thought and practice which had gone beyond the tentatives, the patient restraints, the quiet waiting and alert passivities of Seekers—such as the Therapeutæ—and become men who had found a special initiative, an active procedure, a definite knowledge? Did they know how and why they produced definite results, and were those results such as we might expect from an enlarged consciousness? Had intelligent, planned asceticism and an ordered, purposed life led to a real higher knowledge, a Gnosis, and had this Gnosis in turn produced its appropriate power, power over the mind, over the body and over circumstances?

There is reason and evidence to suppose that this was so. The first stage, power over the mind, is shown by the startling sense of calm which visitors reported as pervading the place and coming from all the members. The community did create a mental "field" which could be felt. The second stage, power over the body, is shown in the most evident way, a way which in Hatha Yoga, the Yoga of physical fitness, is considered decisive. It was remarked how customary it was for Essenes to live to far greater ages than those outside the order, although the surrounding population was of that agricultural and pastoral Semitic stock which had a high record of longevity. Perhaps even more remarkable to our hyper-sensitive age is the testimony that they could and did render themselves wholly superior to fear and pain. Here is clear evidence of some psychic method or technique whereby a voluntary control had been obtained over reactions normally subconscious. Longevity might be dismissed as due to a naturally long-lived stock adding to an hereditary endowment the advantages of freedom from worry and strain, regular exercise and rest, and a frugal diet. The power to control fear and even more, pain, is no happy accident but due to deliberate exercise whereby the threshold of consciousness is pushed back until the will has a firm and intelligent hold on the central nervous system. Nor is this claim an unsubstantiated assertion. During the frightful campaign when the Roman army destroyed Jerusalem, the Essenes were on several occasions captured and tortured to make them disclose their secrets. It is recorded that not only did they endure the torture with-out yielding but that they suffered it cheerfully. They were masters of that technique whereby the energy, which does, in an undeveloped and still divided consciousness, show itself in pain, when the body is injured, can, when the mind is developed, be deflected so that no "raw" sensation remains to be exploited.

So much for the evidence that the Essenes had attained to direct power over the body as well as over the mind. Had they reached the third and final term of power over circumstances? Had they so far advanced the focus of consciousness and expanded the aperture of awareness that they were, at least at hours of enlightenment, able to attain such open vision that they could transcend time? Josephus specifically mentions three of their master seers, by name: Judas,[2] Menahem[3] and Simon.[4] All these three, he maintained, demonstrated prophetic powers, foretelling future events. We can, therefore, assume that the simple asceticism did not account for their development but was only a preliminary training and that, as is frequently the case in these pioneer attempts at an advanced psychology, under an anthropomorphic terminology—in this case the secret names of the angels—was concealed the description of psychological knowledge and the instructions and training as to how to win and use these profound controls. Certainly we know that, under the phraseology of alchemy and astrology, more than once was disguised knowledge of the endocrine system and how, by breath control and concentration, to release hormonic charges which not only changed the energies and powers of the body but profoundly altered the qual-

2. Antiquities xiii, 2, p. 2.
3. *Ibid.*, xv, 10, p.5.
4. *Ibid.*, xvii, 13, p.3.

ity and extent of consciousness; while in most Yogic systems the ductless glands are referred to as Centres of Light or as Lotus Flowers and the basic spinal ganglia is called the Coiled Serpent or the Lady Kundalini.

An order which had attained such specific knowledge and reduced it to a system, might have been expected to make a permanent contribution to civilization, to raise society beyond wars and revolution, to be an irrevocable step in evolution. These pioneers, having so firmly occupied the new line, the frontier of man's understanding and power need never be withdrawn. There need be no retreat from this level—the level of an organic society, which could train the individual so that the enlightened mind directed a controlled body, created a social order, so that there was economic justice and balanced supply of all necessary means, and produced leaders whose power and vision would convince the immature of Life's real meaning and sanction right conduct throughout mankind. Yet the Essene order-vanished and left no visible successor. The reason for its disappearance we shall probably never know. We have, however, a few hints which indicate that the development of the Order was not sufficiently advanced before the Roman invasion, sweeping over their retreat, shattered what was still an experiment and not a completed technique, a complete scientific way of living in and through enlarged consciousness. We know that they seem to have retained in their discipline one obvious rudiment of barbarity. Blasphemy was punishable by death. That such a penalty should stand in their code—whether it was enforced or not—indicates two things: In the first place, it is clear that they had not broken with their anthropomorphic past, for they still felt that there was a deity who was so petulant that he could be offended by insult and soothed by being offered the body of his insulter. In the second place (and this is even more significant of their experimental condition), they had not established beyond fail the psychological ascendancy of their proficients, for had these attained to a complete and demonstrable authority they would not merely never have had to punish, but the wish to violate the rules of the society would never have arisen. The violator would have felt checked before his violation; the sanctioner would have known that if punishment can be called for, then all the community, the sanctioner included, are guilty and merit penalty; for it is rightly taught in India that the guru is so morally responsible for his chela (his disciple in training) that all the chela's faults and mistakes are actually the guru's.

This blasphemy punishment is indeed proof that they had not attained the corporate unity which, for example, a traditionally subconsciously-ruled society such as the aboriginal Arunta tribe of central Australia were still sustaining when they were discovered by the Whites. For in these early—or arrested—communities, although the whole of life was limited, shaped and controlled by tabus, there were no penalties for the breaking of them, because, by the very breach, the individual would cut himself off from his essential life, the life of the whole tribe.

There is also another fragment of evidence which goes to show that their methods, though advanced, were not complete and infallible. This is not from their rules, which might not have been enforced, but from their actual history. Josephus records that, in the desperate resistance opposed to the Romans under Vespasian and Titus, one of the most resourceful and courageous fighters was a certain John the Essene.[5] Such a reference recalls to mind the tragedy of Zen Buddhism in Japan. This, also, began as one of the severest and most efficacious methods of training the psyche to transcend the body, the self and Time, so that consciousness might attain to complete and lasting expansion (Illumination) . It has largely sunk to being a method of mind training whereby the individual may attain to a higher self-control for any purpose, to serve the imperialistic delusions of his country or even to make himself a more efficient business man. The case of John the Essene seems that of one—and perhaps the one is representative of many—who learnt from the Essene training just enough to make him a cooler fighter, a more implacable and dangerous enemy.

After this pioneer attempt the Jewish genius retreated. The Pharisaic degenerative development asserted itself—the belief that strict

5. Jewish Wars ii, 2o, p. 4; iii, 2, p. 1.

traditional observance and pure racial descent assured success in this life and a similar success in a future life after a bodily resurrection. Psychology relapsed into heredity. Religion, which had almost attained full spirituality became re-nationalized. The Chosen people, the Holy Land, the traditional Scriptures; the Seed of Abraham, the sepulchres of the Patriarchs, the site of the Temple, the sacred and unalterable letter, these were the realities. Man's spirit had no future. Open vision was blasphemous. Light only shone from an ever-receding Past. For a short time in Alexandria (where as we have seen had lived the Platonizing Jew, Philo and his Therapeutæ, "genuine mystics of the monastic type,"[6] an order of elderly men who in late life achieved innocence but not power, scholarship and gentle saintliness but not open vision), the Synagogue allowed, or did not actually condemn speculation and, no doubt, also some practices somewhat more psychological and scientific than the orthodox service of reading the Torah. Egypt was always a place of spiritual pioneering (as we have seen and shall see when mentioning the monastic experiments in the Thebaid). That the Jews there were more independent of centralizing tradition we know not only because at Elephanta and there alone throughout the world, did the "Dispersed People" actually build themselves a Second Temple, but also because there free thought continued. We know, if only by its condemnation, that speculative thought, which might have freed them of their racial obsession, went on till the sixth century A.D. at Alexandria because it was at that date that the Synagogue there finally forbade to orthodox Jews the right to believe in reincarnation. Condemned, but never wholly suppressed, the Jewish Cabbalists—Hebrew thinkers striving to find the narrow path out of a dead legalism between the fogs of unsubstantiated speculation and the sloughs of magical practices—attempted to achieve a real psychology, a secret method of dilating consciousness. Their attempt, however, does not belong to this chapter, for it was in the Middle Ages that they made their principal effort and it was from the Renaissance down to the early nineteenth century, from Pico to Blake that their

6. Inge, *Christian Mysticism*, p. 82.

influence—a stimulant to seek and experiment, rather than a guide—is felt moving behind the scenes in Europe.

So closes the specific Hebrew attempt to explore and expand the limits of consciousness. The monotheism, which was the outstanding characteristic of this branch of religious development, though it has been thought to be a sublime achievement, really arrested psychological evolution. It must be repeated: Once man conceived that the unity of experience was in reality divided into three fundamentally separate entities, Man, Nature and God, he had entered on a blind alley. All his attention henceforward would be confined to seeking for ways and means whereby God might be persuaded to alter Nature to suit Man. The idea that outer Nature was a misconstruction (Maya), a misapprehension, became ridiculous. Whatever man saw outside himself was hard, inescapable fact. The senses were not to be mistrusted as bad witnesses. Through them man directly apprehended reality and it was alien. The idea that an external, anthropomorphic God was an abstraction became blasphemous. All that man could do was to pay his deity such honours and tributes as would purchase his interference. The true path of evolution, through expansion of consciousness, had, therefore, to be abandoned. Hebrew religion no longer strove to understand those inner urges and convictions which assured it that there was a God, or attempted to inquire into the nature of its apprehension of the outer world. The prophet therefore becomes completely extinct—the Temple is dismantled. Seers are no longer acceptable, for what can they see beyond what everyone knows, the two unalterable realities—a mechanic Nature and a magnified, non-natural Man? The Temple ritual, which was once a believed magic, is also now unnecessary. The minute observance of the Law, and through it the fossilized perpetuation of the Holy people, this was all there was to do and all there was to hope. The magnificent attempt of Hebrew genius sinks back into a racialism which later societies have found all too easy to resent and also to imitate. Failing to sustain the evolutionary advance and to emerge into a higher phase of consciousness in which a new, objective order would be perceptible, and the individual transcended in a new state of being, the Jewish spirit, tormented by that unresolved individualism, could only seek to assuage itself by relapse into a nationalism which would partly reabsorb and partly inflame its egotism.

Such a development bears all the marks of evolutionary degeneracy. Now that we are seeing this development take place, no longer among a small pioneer people, but in the master races of the world who are free to follow to the full the dictates of such an impulse, we can have no doubt that this is decadence which if not arrested must end in the self-destruction of the species.

CHAPTER VIII

THE HISTORICAL EVIDENCE FOR THE EVOLUTION OF CONSCIOUSNESS

III—Greek

WE HAVE now reached the third of the sources from which our religion and psychology arise—the Greek source. It is nearer to us both in place and time and so can rightly be put after the other two. We have seen that the Hebrew carried on, both into the clearer light of history and also to a further stage of actual religious explicitness and definition, the psychological evolution begun in Egypt. In turn, now with the Greek, we shall see that evolution, which became arrested in the over-defined, over-projected deism of the Hebrew, carried a stage further. The Greek attempts an experimentalism, a scientific approach which is an essential stage in the growth of awareness, but which the Egyptian would probably have looked upon with the contempt with which an artist regards a book giving rules for painting a masterpiece, and the Hebrew would certainly have dismissed as blasphemous.

Thales, the founder of speculation and empiricism, has already by the sixth century B.C. arrived at the division of the world into spectator and spectacle, subjective and objective, an outer, single-substanced, uniform and regular Nature, and an inward, conscious, clear-seeing, detached Mind. This is the unscientific assumption that the outer world as directly apprehended by the senses (i.e. by uncorrected unexpanded consciousness) *is* reality, an assumption which is ruling our present civilization. That this assumption was recognized as becoming rapidly popular and dangerous is seen by Heraclitus (540—475 B.C.) at Ephesus, protesting and warning that the senses are "bad witnesses" and that only the wise man who has trained his consciousness can have true and accurate knowledge. That this life without such a scientific discipline is a state of drugged hallucination, he expressed in his phrase that while we live, our souls are dead in us, and when we die our souls are restored to life. His protest was, however, insufficient to alter the disastrous widening fissure in thought, for the mistake in rudimentary science was due to an earlier one in theology. The conscience and higher consciousness of man became, owing to his growing individualism, projected and (as always) anthropomorphism resulted. Finally these Gods, who are merely human passions given separate bodies, are found to be too disreputable. The reform of Euhemerus (300 B.C.) makes Olympus respectable, but the bowdlerization of folklore removes the Olympians still one more stage from reality. Hypostatized passions could have some indirect bearing on actual living. Appealing to Juno, a matron might rouse her latent maternalism. A vow to Mars might make a hesitant fighter brave, one to Venus or Priapus urge a feeble passion. The force, however, would tend only to increase these primitive desires; the suggestion being the more powerful the closer it came to animal craving. Render all the Gods faintly noble and their aid would become increasingly ineffective.

This fissure, which if let spread must have proved fatal and brought Greek civilization to an end, was, however, arrested though not healed. Into a religion already advanced in pernicious anemia of the spirit, came two forces, or a force with two aspects—the one which could restore men's experience, and the other their knowledge, of a more actual, more intimate and more extensive reality. The first were the invasions of the Bacchic rites which reached their climax before the sixth century B.C. Dionysus is first heard of, in Homer, not as a God of wine but as a wild Thracian deity who could produce "raging," the prostrating de-tensioning, a semi-epileptic frenzy resulting from that utter freedom of behaving regardless of consequences.

It may be asked how did such a "decadent" religion manage to enter Greece as early as the Homeric age, when, whatever men might be in the Heroic epoch, they were not doping and soaking addicts. They drank in the evening and in bereavement took soporifics (see Chapter V), but their everyday activities surely show them as men who never doubted that the life of action was always worth living. It may be worth asking the question whether the attention given in Homeric times to a religion of rage did not spring from a newly recognized need for anger. The Homeric heroes are not philosophical speculators but they have reached the first stage of inquiry when man begins to have doubts about himself. He is aware that he is no longer singleminded in his lusts. He begins to be aware of himself and to hesitate. In a fighting world the hesitant Hamlet is lost. As will happen again later, when the Scandinavian Heroic Age recapitulates the Homeric Greek,[1] the man capable of frenzied, unreflective rage—the berserk—will be envied, if not actually admired, because even the bravest fighter is discovering the cold truth expressed in Rochester's cynical line: "For all men would be cowards if they durst." The berserk, as Bartlett in his essay on berserkism pointed out (and the Gadara skull found in the Greenland medieval extinct settlement of that name, confirmed) was a man whose physique and psyche had undergone a degenerative change through failure of the pituitary gland. He suffered, therefore, from frenzies of rage (succeeding long periods of brooding). During those rages he was oblivious of danger and fought without armour (hence ber-serk—bare skin) and, owing to the immense charge of adrenalin released in his blood by the excited pituitary, working on the suprarenals —his energy was prodigious, he was incapable of pain and when wounded he did not bleed. No wonder such a creature seemed possessed by a daimon if not by a God, and no wonder also that those more advanced Heroes who felt the first chill, when "the native hue of resolution is sicklied o'er with the pale cast of thought," longed for some religion of frenzy and raging which would give them a similar ecstasy. Once the reflective fighters, the men of calculated courage, through their need for rage, for what Homer calls "The Anger which rises in a man's mind like smoke and is sweeter in his mouth than honey," had admitted a religion of frenzy, the next stages were inevitable. The ritual which had been used to string a man to the ordeal of battle, like all drugs, began to be taken to render everyday living endurable. The war-dance induced, inner glandular storm had to be supplemented by alcohol,[2] and Dionysus becomes a wine god.

---

1. See Chadwick, *The Heroic Age*.

So a series of such transfusions of orgiastic energy entered Hellas from Thrace. This countryside was no doubt one of those zones of psychic disturbance which seem usually to form in areas which lie between regions where a simple, unreflective traditional life of action is still being lived, and districts where (in this case Hellas) thought and speculation have already arrived at the first antinomies, the first baffling puzzles as to seer and seen. From Thrace, therefore, were to come a succession of convulsive enthusiasms ending with the Orphic Mysteries. With the rise of this mollified ecstasy, that centre of spiritual cyclonic energy died away, and it is worth noting that as civilization passes north into Macedonia, pushing back the simpler peoples of unreflective action and unhindered upflow from the subconscious, it is from the Thracian border town of Stagira that comes the last and greatest of Greek analytic minds, Aristotle. The unconscious creature of action, passes into the strangulated consciousness which strives to relieve its suffocation by epileptic efforts and ends, "all passion spent" in the great "Master of the Mean": the Mean, that balanced life which no longer aims at further enlargement of consciousness, the further evolution of the psyche, but desires only a deliberately stabilized and arrested state, a state when man concludes that as he is and as he knows, he could be happy.

The Bacchic crusades were efforts to win the Greeks back from a life of sophistication and abstraction to a life of emotional reunion. By drunkenness the self-conscious man was able to sink again under the *limen* into the subconscious. Dissociation was induced by alcohol and dancing. Sexual stimulation was sought for the same purpose, and at the same time, for Herodotus notes that the phallus and the ithyphallic image, carried in the Bacchic procession, were in his time a specific part of the Dionysan worship.

Such a religious excursion could lead to no satisfying goal. Evolution is irreversible and the only solution of man's problem of psyche conflict, as of physical pain, is to use the pent and torturing energy for an enlarged intensity of consciousness. He cannot return to ani-

2. The Homeric Heroes of course drank heavily, but in the evening, *after* fighting, not, as does the later alcoholic, before fighting, to gain "Dutch Courage" and finally before facing any day.

mal awareness or unawareness. The fact that he has to use alcohol to delude himself that he has so slipped out from under his problem is proof that no return to an unthinking healthy animalism is possible for him. That return is barred not merely by his own nature, which still has stores of unused energy (his torment and his hope) but by the nature of all contemporary animalism. There is, in fact, no healthy animalism into which to turn, for, as we have seen, all the animals are already living fossils. Man alone is still fully alive, still not decadent, though tempted by his torture to desert life for death. Dionysanism does show the bankruptcy of the anthropomorphic Olympian religion. It reveals the fact that any inner impulse was better than these cold abstractions. The Greeks, contrary to Wordsworth's romantic notion, would rather have one inner intimation of immortality, even though it were only liquorous, than see any number of animated Elgin marbles crashing through the undergrowth and polluting the water supply. It was not enough to be a pagan suckled in a creed outworn. They were ready to be sots, fanatics and indeed maniacs, if only they might lose themselves once more in a vast sea or storm of undifferentiated being. The frantic effort ended, as it had to end, in criminal extravagance. Dionysus Zagreus, one of the forms of the religion, was worshipped by tearing in pieces a live boy and devouring his flesh raw. The cruelest aberrations of asceticism scarcely outstrip such a horror. Yet from this we can learn how immense is the raw energy pent in man and how, if it is denied creative expression, it will destroy both the mind which fails to find that expression, and also the body which attempts to give it outlet.

That even such bewildered people as the worshippers of Dionysus realized dimly what was their real need—enlarged consciousness—and that they actually had hoped to win this open vision through drunkenness is shown by the curious fact that Dionysus was said to be buried at Delphi, Delphi the home of the greatest of oracles, the place where Greeks were to make the greatest effort to extend their knowledge of reality by an approach to a scientific psychical research. Indeed, at one time the Dionysan oracle—the attempt to attain extra-sensory knowledge by means of alcoholic intoxication—

rivalled Apollo's.

Why the Apollonian method won against the Bacchic, and how it was to turn Greek thought from the vain attempt to restore an animal experience, to the "Right Effort" to attain to widened consciousness, to the apprehension of an extra-sensory reality, we must now see.

The Apollonian oracular development, although it also failed, was both a progressive (instead of a reactionary) movement in evolution, and also one unstained by horror and violence. This may be called the true mediumistic method. For here, in contradistinction from the way of frenzy, the medium falls into a trance resembling sleep and while in that state pours out remarks, or when questioned gives answers which the attending onlookers record. The attempt to make Dionysus an oracle is no doubt a transitional phase between group frenzy, in which all felt transported but none brought back any knowledge, and developed mediumship, when the feelings of the medium, however remarkable to herself, are irrelevant and the information which she utters (and of which she is nearly always ignorant) is what is of value. The aim of oracularism is now to get beyond any physical seizure or ecstasy, however wonderful, however, even, cathartic. Men feel that it is not enough only to attain a state in which the acute Problem of Life will be submerged and forgotten. The objective of the new method is to bring back extra-sensory knowledge, actual apposite information, in the light of which the problem, which had defeated all rational knowledge, may actually be solved.

In this context it is worth noting that according to the Greek Orthodox Church, mystical experience must depend for its authenticity not on the intensity of the Mystic's feeling of conviction, but on the nature of the information brought back.

The first stages of the oracular method *per se* are visible not at Delphi, however, but much farther north, in Epirus. The most ancient of the great Greek oracles is that among the oak-tree groves of Dodona, in the north-west. Whether this was a shrine of the oak-worshipping Dorians (who saw Zeus the Thunderer marking his presence in the lightning shattered tree) or whether they destroyed an earlier culture which left this, a last fragment of its religion, we know from Herodotus that the oracle had a greater psychological interest, than had it been no more than a way of guessing the future by arbitrary association between the marks made on the oak by electricity, and human events. Even this early and ruined system made use of mediumship. The oracles were delivered by priestesses, he records, women who, he adds, were dark and, he further speculates, of another race. He believes that they were foreign. It is far more probable, however, that they were survivors of the original brunette Mediterranean population, who looked almost Egyptian among the blond Dorians now settled throughout north-west Greece. These people, especially among their females, would very likely retain a gift for dissociation and even extra-sensory perception. As the same dark race to-day produces such types, who impress the conquering Nordic with the exercise of powers which he lacks, so the Dorian Nordic may have spared these women to fill a similar gap in his armour against fate. They may also have retained, from their perished society, some rudimentary methods for dilating consciousness. If these women were of the medium type, the soothing murmur of the wind in the thick oak leafage would not be what gave them their oracular information, but it would act as that distractant of the surface mind's attention, a distractant which it seems is nearly always essential, if the subconscious is to rise to the surface and control the speech centres *(vide* Elisha's technique, Chapter VII). That their surface mind may have believed, as did their blond masters, that the murmur was a whisper of their Father God, would in no wise interfere with the functioning of their misunderstood faculty: quite the contrary. The surface, individualized consciousness is the less inclined to inhibit uprushes from the subconscious, the more it believes that what is speaking is an external individual like itself.

The Dodona oracle lies, however, at the limit of our historical vision and what we are told of it would seem to show that it was rudimentary. We may, therefore, turn to an oracle which was predominantly famous throughout the brilliantly lit period of Greek history, an oracle of which we have ample records and which impressed the best minds of the ancient world. Rich and cunning Delphi undoubtedly became, but behind this accumulation of tissue, what was the insight which raised it to a position of unique authority? The competition in oracles was keen. The public of the ancient world was not confined to one soothsaying establishment. Before Delphi could afford to pay for its private detective service it had to build up capital, and that capital came from the gifts of those who had found Apollo's indications uncannily perspicacious.

First, it is probably instructive to note that the God who here was supposed to give illumination was Apollo. Nietzsche made the division of all religions into Dionysan and Apollonian. He intended this to mean no more than human beings tend to delude themselves in one or other of two ways; either by orgiastic violence or by seemly drill. His distinction, however, is probably deeper and more valid than he suspected. We have seen that the Dionysan religion was an attempt to throw back into a pre-individual state of mind, and it too attempted oracularism, through alcoholic frenzy. The Apollonian is a far more scientific effort to climb up into a mental condition which may be post-individual, superconscious. So as Dionysanism sank into exhaustion, at the other end of the religious spectrum, through the seemly worship of Apollo, another effort was made, an effort, deliberate and experimental, to discover a technique whereby consciousness might be dilated. This technique employed, as did the Dionysan ritual, physical disturbance, a toxic condition, so that while the common senses were distracted, the psyche might shift the hoodwink of its body and obtain insights of a super-animal world. The development of the oracle at Delphi is the Apollonian, the daylight, scientific attempt to reach out into a larger world of reference than that given to us through the senses. It, therefore, attempts to lay hold of this higher reality with a quiet and steady grasp, as a climber stretches to a further handhold, taking care not to shift his weight and leave his present support until he is fully certain that his purchase is sure. In contrast with this approach the Dionysan effort is a leap—a violent attempt to fling the psyche out of the body, regardless of where it may fall.

Nor was the Delphic experiment scientific only in its patience and steadiness. It was even more so in its employment of researcher and subject—in dividing up its forces when attempting this exploration. The problem of dilating consciousness presents always a great difficulty. It is, in fact, a double problem. The first part is to attain to a radical extension of awareness, to see clearly a world other than that in which we function in common with all other animals, except—and this is a vital distinction which makes their continuation possible and ours so precarious—when they are guided by race memory, by instinct. The second part is not less difficult; it is to bring back to the present normal, analytic consciousness the suprasensuous data obtained in a condition when the analytic consciousness was dissociated. The scientifically-minded Apollonian priests at Delphi attempted to solve this problem by what appears to have been a careful development of mediumship. They selected from the valley which lies under the Pythian shrine, women who easily passed into trance. It is not likely that the existence in that locality of a biological strain, carrying a strong faculty for extra-sensory perception, decided that a shrine of Apollo should be at that spot, but there is increasing reason to suppose that to this hereditary strain, and the priests' discovery of how to exploit this faculty, is due to the complete supremacy of Delphi, not only over all other shrines of Apollo, but over every other oracle of the Hellenic world.

We have seen that there is evidence that, at earlier Dodona, it was the combination of a famous site, the often thunder-stricken oaken grove, with mediums who thus could get a hearing, which made that place famous, though there was no priesthood to train the women. They were priestess-mediums in one. Delphi was, then, a step toward a definite scientific technique. That this Delphic priesthood had, however, to come to their knowledge gradually, is shown by their dependence at the beginning on the carbon dioxide fumes which escaped from a fissure in the Sanctuary floor. We know that they made the Pythoness medium breathe this vapour, and were so firmly convinced that in it resided the inspiration, that when one vent ceased to discharge they moved her tripod seat to where there was another escape of the gas. It was, however, when these earth fumes finally ceased that the priesthood showed its enterprise, making the medium lose surface consciousness through inhaling the smoke given off by burning laurel leaves. By this time they may have become aware that the medium was more important than the method, that it was the actual Pythoness rather than Apollo who possessed prophecy.

That they clung to the dangerous technique of bringing on trance by partial asphyxiation does not necessarily prove their superstitious ignorance. All modern research with mediums shows that few can function unless they believe that they are the instrument of a higher Being. Why this should be so has been suggested above, when noticing the association of medium priestesses with the Dodona oak grove. It has also been shown that any disturbance of the technique to which she is accustomed almost invariably inhibits the manifestation of a medium's gift. What is of more gravity is the fact that while the priesthood employed an experimental subject whose consciousness they might dilate, they themselves remained as critical onlookers and listeners so as to collect and order the random utterances the medium poured out while in trance. Such a technique is certainly best if what is desired is extra-sensory perception, knowledge of future events which may be exploited for personal gain. If, however, the aim is for each and all to attain to a continual experience of supra-sensuous awareness (or at the least, for the loftiest minds to have such experience) then to do no more than to send across the threshold an inferior intelligence and rest content to listen to what it mutters in its dull amazement and through its drugged brain, is to repeat the fairy-story tragedy of one, who, offered a kingdom, came away doltishly content with a couple of gilded farthings.

The Delphic priests were right in insisting on obtaining real information applicable to the problems of the actual world and in submitting the mediumistic utterances to critical and analytical examination. They were wrong in resting content with nothing more than such soothsaying; in failing themselves to explore that world of enlarged consciousness into which the medium peered in trance. Had they themselves also attempted to understand her experience and even to take part in it, they might have won to a real, new open-vision; and so instead of their college being content only to exploit pre-vision, by telling rich men what was coming to them, they might have begun to understand what is the true nature of that misapprehension which we call Time, a misapprehension which deforms our actions and renders us subject to, instead of master of, circumstances.

We need hardly wonder then at the degeneration of this oracle, and that a priestly college, which, at the height of its power, gave Greece the Amphictyonic Council—a courageous attempt at a Hellenic League of Nations—gave the world the conception of Sophrosyny (that organic balance of all the virtues), and wrote boldly over their temple the supreme oracle "Know Thyself," through failing in precisely that essential knowledge, sank to the level of covetous charlatans. It was because the priesthood itself only knew at second hand, and in a degraded form, the larger world of reality, that the great experiment at Delphi has remained no more than a psychological puzzle. Lacking the leadership which such a priesthood (had it developed, at first hand, dilation of consciousness, and itself attained enlightenment) could have given, the Greek genius could only consume itself. Before, however, we leave this experimental effort we should distinguish quite clearly its double achievement. In the first place, the Delphic priesthood had the moral and truly scientific courage to perceive, in a civilization which was rapidly becoming self-conscious and analytic, that a psychic factor, essential to the moral order and the sanity of mankind, was being lost among their own educated classes and now was only lurking among simple women in an upland valley. They recognized that a dilation of consciousness ("open vision"—to use the Hebrew phrase) was essential to society, was practically lost by the educated—the very people who were shaping the civilized world in their own image—and now was only retained by poor and rudimentary types. In the second place, they reasoned that some regular system or procedure must be devised if the spontaneous, sporadic emergence of open vision among persons of low intelligence was to be made to function under control and yield reliable results. They began to devise methods whereby dilation of consciousness might be brought on at command.

Though, then, they failed to solve humanity's evolutionary problem, the problem of achieving supra-consciousness, they stated it with a new clarity. Though they failed, other seekers elsewhere were to push the exploration further.

We may then, perhaps, correlate the Delphic experiment with that phase in Egyptian religious development when the re-established and reformed religion of Amen began to provide esoteric initiation and training for those individuals who were sincerely anxious to advance. Both mark a time when the rank and file are beginning to be left behind. That time is a third epoch in religious evolution, the first being the period of organic community, when society advanced slowly but wholly, and the second being the period of the despot state when open vision has vanished and everyone is held and ground between an iron militarism and a petrified religion. But whereas Egypt was able, through the official religion, to create an esoteric system which allowed the individual to escape (and so preserved the fossilized community from his explosive force), Greek official religion failed to make such a channel. The Delphic worship failed to provide initiation for seekers after enlightenment, hence Greek society could not fossilize as did Egyptian, and was increasingly subject to revolutionary violence. Further, when at length a way of enlightenment was discovered, it was not through the Olympian theology, psychologically reinterpreted, but through a throwback to the Chthonic fertility theology of the prehistoric people—the original Mediterranean stock.

This fact, that the throwback in Greek religious evolution had to be further than the throwback in Egypt's, had serious social consequences. For the religion of Amen, whose esoteric training satisfied, at least for a time, the pioneering Egyptian individual, was the State religion. The community and the individual were therefore not in open conflict: the doctrine of the higher life (for the pioneer) and the lower (for the routineer) was framed as a compromise. The individual did not desert society: he was given leave of absence, or perhaps it would be more exact to say he was "seconded," sent from ordinary social occupations of citizenship to undertake a particular task. The Hellenic granulation of the originally homogeneous society had gone further. Therefore neither could the Greek's religion develop a psychology—it had to remain fatally fossilized in an irrevocable anthropomorphism—nor could his own individualization—so much denser was the threshold that separated him from his subconscious than the Egyptians—be resolved by any religion with which he was familiar. As his sibyls had to be gathered from the conquered Mediterranean race, so when he himself was to find a way back to the living stem of life, he had to go back to these, his despised, crushed and almost buried forerunners. From them, with helps from Egypt, he was to make his own method of enlightenment. It was, however, to be so alien from the established religion that the success of the Mysteries was fatal to the Olympians.

In Greece we find no attempt to develop an intentional socio-economic pattern. In Palestine and in Egypt, because the higher religion had sprung directly from the lower, and Amen and Jahweh were as much Gods of the esoteric initiate as of the exoteric conventionalist, even when pioneers found it necessary to advance faster than the loftiest members of the priesthood could permit, these pioneers did not develop as solitary secessionists. They founded patterned communities for avowed living. In Greece, because the Gods of the initiate were not fully the Gods of his fellow citizens, there is lacking this middle step between esoteric initiates still attached to the State religion and those solitaries[3] who no longer have any hope of mankind's evolution on the physical plane.

We must now trace very briefly the steps (or perhaps one should say the leap) whereby the Greek arrived at that final term, that stage when he attains enlightenment, discovers that this is only for the few and has to decide what is his duty to those behind: Is he to remain trying to shame them into taking a step they do not desire, or is it his social duty to go on, attain complete purchase on and station in the wider and deeper reality, and then, once his fulcrum is firm and below the flux of Time, fundamentally affect the future of mankind?

---

3. e.g. Apollonius of Tyana, the Neo-Pythagorean contemporary of Christ. It seems clear from his life as given by Philostratus that he underwent Yogic initiation in India and attained to enlightenment sufficient to give him complete awareness of the nature of Time, for two of his remarks show his aim and the level of attainment which he knew himself to have reached. "All men wish (!) to attain to the constant awareness of God but only the Indians achieve it." "I claim no more than the power to foresee the future."

We must recall that the problem which has confronted man during history is how he may forward his evolution. He has unused still within him such pent vital energy that he can still evolve. His evolution, however, is no longer physical but psychical. If then he is to evolve, it will be through an extension of consciousness. His development demands such an extension because the world, as he now sees it with sharpened animal senses, gives him no sanction for using well his immensely heightened animal powers. The guards of instinct have been removed, but the guidance of vision has not been gained. The task the Greek attempted, as had the Hebrew and Egyptian before him, was to find how to release creatively the evolutionary energy without shattering the body or unhinging the mind.

The Mystery Religion phase in the psychological history of Greece is better recorded than is the parallel phase in the psychological histories of Palestine and Egypt. As we have seen, we have for this epoch in Egypt only the Hermetic writings and perhaps Philo's account of the Therapeutæ, and for the corresponding epoch in Palestine, the scanty record of the Essenes. Nevertheless, the history of the Greek Mystery Religions is very ill known to us and is likely so to remain. Such clues, however, as we can collect yield us some important pieces of information. First, we must note that as the developments of Egyptian, Hebrew and Greek religion overlap, the Hebrew starting after, and in some ways below, contemporary Egyptian religion, and the Greek repeating this recapitulary development, so in Greek religion itself, one stage will still be developing when its successor is already springing up.

Discoveries in south Italy would seem to show that the Pythagorean order, which was dominant in the sixth century, was not only part of that Eurasian individualistic revolution which swept away the fertility religions, but that the ideas which the Pythagoreans expressed were imported. The famous golden tablets, dug up in south Italy (these have been called Orphic) use such phrases as "I have escaped from the wheel." There is also the tombstone inscription which runs, "freed from the grievous circle" (of births and deaths). These can hardly be other than references to the Buddhist doctrine of rebirth. The whole Pythagorean effort, its dominance as an order over the social life of such City States as Croton, and its sudden destruction by revolution, suggests that here are foreign and maybe Indian ideas, such as reincarnation and caste, being prematurely imposed on a people who themselves had not reached that stage of thought and conviction.

The Orphic order and mysteries are less clearly mentioned. Here, too, however, we seem to perceive an alien notion, a sudden development in Thrace, ahead of native belief, introduced by a zealous leader. Certainly Orphism taught "We are here in a kind of prison" and believed in transmigration. Under the tale of the magic singer who "drew iron tears down Pluto's cheek" we may have reference to some "sound Yoga," an enlargement of consciousness brought on by music. What seems equally clear from the Orphic stories is that the movement failed and its leader was killed, perhaps, as Jane Harrison supposed, because he tried to teach the Thracians a higher ecstasy than alcohol. The movement continued as a specific Mystery Religion; Dionysius Nyctelius records that the initiates met at night. The movement, as Herodotus notes, is one which approaches to Pythagoreanism, though whereas Pythagoreanism became a political movement, a psychiatry, an economy and a policy, Orphism remained only at the first stage. It is a transitional stage between Dionysanism—the religion of ecstasy by intoxication, bouts of world-forgetting frenzy—and a complete, constant, intentional way of living. From it spring those religions of private salvation which we detect under the specific name of the Mysteries.[4] Certainly after what we may perhaps call these premature efforts (and after the Delphic priesthhood had shown that dilation of consciousness was both necessary and possible—necessary for social order's sanction, possible through trained mediums), the Mystery Religion takes its place as the next step in this exploratory evolution. And as there were many oracles, sporadic and natural, but Delphi became supreme, because it strove to make the faculty work under control and subjected raw out-

4. That Indian thought was known after Alexander we know if only by the fact that Pyrrho, the founder of Scepticism, went to India with Alexander's armies and, having been trained by Brahmins there, returned to Elis, his home.

crops to critical analysis, so, too, there were many mysteries (beside the Orphic), but the Eleusinian became dominant. The Eleusinian shared also with the Delphic experiments the fact that neither was foreign, each, when it had to fall back and "spring from below the graft," sprouted from the buried subconscious, sprang from native traditions. Neither was a foreign importation, as it would seem with the Pythagorean, the Orphic and the Dionysan. At Delphi the sun god Apollo—an extremely externalized sky or Olympian anthropomorph—is recombined with the fertility, earth or Chthonic Python he slew. The original Chthonic religion—a primitive snake worship—reestablishes itself as the effective element. The priestess-oracle huddles herself over the crack out of which the snake of wisdom was supposed to have crawled, sitting on a tripod made of three entwined snakes, and she herself is called after the great snake's name, the Pythoness.

So, too, the Eleusinian rite was based on a Nature tradition but in this case the throwback had to be further. At Delphi the sun god is made into a sleeping partner. He is shelved gently, and while it is the ancient religion of the conquered which gives the machinery of inspiration, the first name in the firm's title remains Apollo. At Eleusis no blue-blooded fainéant Olympian introduces the dark, effective deities to their new, propertied worshippers. These deities are the ancient earth goddess Demeter and her daughter, queen of the Underworld, Proserpine.

The Olympian Gods, like Akhnaton's Aten, were projections completely divorced from the fertile subconscious mind. They were become gods, wholly other from man. As soon as that happens religion becomes for mankind seemly, pensioned and untrue. Anthropomorphism appears to promise to make deity human, but because it is only a projection of man's sundered fore-conscious, it can only end by making religion unbelievable. Though, however, the Mystery Religions had to go back to the Pelasgian goddesses—the simple fertility deities of the peasantry—those who effected this revival avoided two mistakes and so brought religion a step further. The first was the Delphi mistake of being content with nothing more than secondhand knowledge of extra-sensory perception. They realized that if society was to be saved from capsizing, from analytic knowledge completely outbalancing integral insight, it was not enough to have mediums who could in certain circumstances occasionally be capable of telepathy and clairvoyance. The rulers might use such information and even be prepared to pay for it: their conduct would continue to be not less deadly to civilization and humanity. No more than high explosive has made power politics more humane, can second sight make Machiavellianism less crafty and more enlightened. What was needed was that everyone who chose to undergo training, might have direct experience of the larger reality which alone can sanction morality, and that sufficient numbers of high-minded men might have this experience, so that the world might by their example at least be restrained from self-destruction if not inspired to new vision and new being.

The second mistake, successfully avoided, was to attempt no more than a simple revival of the old faith. This is an all too common reaction among advanced analytic people. When they find that their own line of advance has given out and that their powers are not bringing them profound understanding and unity, they are apt to rush back incontinently and to believe that only among savages does integral wisdom still reside, and only in a blind acceptance of the simpleton will sanity be recovered. This is neither possible nor true. The ancient Pelasgian religion did not in itself provide an adequate vehicle for further psychological exploration. The masters of the Mystery Religions grasped this vital fact. Those individual Greeks who realized the Olympian Gods were lifeless abstractions, but still sensed that religion, though dead at the top, must yet have life in the roots, were ready to seek in the rites of a conquered race for such points as might yield a new evolutionary departure. They were prepared also to combine foreign knowledge with native experience. Without some method of procedure worked out by others longer on the task, it is doubtful whether the Hellenic mind could have made a further contribution to the methods of expanding consciousness. What is certain is that in the lower layers of the deposits at the Eleusinian shrine, have been found religious objects which are Egyptian of the eighteenth dynasty. Here, as in so many more manifest subjects,

Egypt was the nursery which produced the seeds from which Greece raised new varieties. We can, then, say that, as the established religion of Amen in Egypt, under attack, was reforming itself, the discoveries which it made in the reinterpretation of its Fertility Myths and ritual were passed on to a part of Greece. For there Hellenic thinkers were puzzling at the same problem, of how, when the sun and sky gods had evaporated like rainless clouds into a hard inhuman heaven, to raise from the dark seeds of an ancient, deposed faith a new and fruitful Tree of Life.

There can be little doubt that this, the dominantly successful of the Mystery Religions, did carry further than before the task of deliberately enlarging consciousness. We know that its success was so wide that Nero, absolute master of all the civilized world, came to receive initiation. We also know that its moral prestige was so high that its priests refused him his request on the clear and clean grounds that no one, Emperor or slave, could be initiated if stained by blood-guiltiness. Unfortunately, we can learn little of the actual procedure and technique. This is to be expected, for the Eleusinian experiment and development belongs, as we have seen, to the individualistic stage of religious and psychological evolution. Further, in Greece this stage probably reaches its acutest point. The Greek, of all the peoples of the ancient world, was the most keenly aware of his separateness, from the world about him (hence his analytic science), from his neighbours (hence his intensely unstable and secessionist democracy), and from his own subconscious (hence his supremely anthropomorphized religion). There is, then, among those religious pioneers who devised the Eleusinian mysteries, an intense native wish to keep their discoveries to themselves. There is no desire to frame a new social order, either by striving to make, through a reform of the national Olympian religion, enlightenment available for all, or by devising some dynamic caste system whereby those who were enlightened might put their vision and authority, as the eyes of an organic community, at the service of all.

We are not discussing here whether this was right or wrong, avoidable or inevitable. We have to note that the stage through which Greek thought was then going, was the stage at which man seeks private salvation. He does not desert the world, but he only serves its lower need for order and physical satisfaction; for stability, not for progress. In the next step he will be compelled by his thought to give up even that secondary service. He will disappear and the line of his advance come to an end. Hence the secrecy imposed on the initiate was never broken, at least by a competent reporter. Surely, we may ask, granted that such a secrecy is an essential part of training, the training mastered, the insight won, was it inherently impossible for any rendering of the vision, any interpretation of the sanction disclosed, to be given to those on the lower rungs of the ladder of awareness? That, indeed, is the supreme question to which all this inquiry leads. Can anything be brought back or must the voyager, of evolutionary necessity, round the cape, disappear from sight, attain a new life and never return?

If there is an evolutionary future for mankind, if humanity is to surmount its present impasse and enter on a new order, is that future for all, through a gradual infiltration of the light from the highest lit to the most dimly aware, or is it for the few who alone can take it in fully? Is evolution to continue with its iron rule, "many called, few chosen," a minute number solving the Sphinx's riddle and the rest flung back into decadence and destruction? Is there to be progress, a pervasive salvation, a steady rise of each grade until all are ready for complete transformation, or do society and civilization end here and now and the few go on to an utterly different condition of consciousness? Can evolution still be psycho-physical, or must it be now, if it is to continue at all, purely psychological? With self-consciousness do we attain the highest degree of consciousness compatible with life? Beyond that must the body become untenable, the blend of soul and flesh unviable?

The Greek Mystery Religions development failed to answer that question decisively one way or the other. For though the mystery technique was kept secretive, its initiates did not fly the world. They remained public-spirited men. Though we cannot exactly estimate the technique, we can judge its efficacy by the lives of its followers. Nearly every eminent Greek of the last great phase of Hellenic culture had been initiated and bore witness by both word and deed to the value of the initiation, while Diodorus says of the Samothracian Mysteries that those who have taken part in them are reported to become more pious, more upright and in every way better than their former selves. That these men were of such commanding intelligence and character (Plato, Aristotle and Plutarch were among them, while the satire of Aristophanes which ate into all other contemporary faith and practice, respected this system) makes us scan with a greater care such details of the procedure as have come through to us. We know there was a sacred meal (a communion) in the first initiation, when the postulant was brought into the presence of Proserpine: for those who proceeded to the second and higher initiation, there was the witnessing of a sacred marriage, when Demeter was revealed. To these two bare facts can be added some details: that after purification the postulant was blindfolded, that if under examination he disclosed unworthiness he might be killed, that he experienced considerable duress and, whether hypnotically or through a pantomime, was given a picture of an after-death condition when, the accidental circumstances of life removed, the soul finds itself in the state it has deserved, and among the conditions its acts have created. The state of mind which the initiate endured and the process of a psychological catharsis, can be reconstructed in outline. The whole experience was summed up as being a termination of ordinary life. We may again quote one of the few initiates who left a description which has survived, though his was apparently Egyptian: "The mind is affected and agitated in initiation as in death: first there are bewilderments, labourings, wanderings, darkness; then horror and trembling. This passes and a divine light displays itself." (Stobæus.)

From these two pieces of evidence, the lives lived by the initiates and the actual character of the initiation itself, we can then see clearly enough why this final Greek experiment in expanding consciousness failed to be decisive. The process itself was too rudimentary. Granted that it was necessary to throw back to the Pelasgian theology and ritual, as the technique evolved was not being used on ignorant peasant women or to cultivate mediumship (as at Delphi), but on the acutest and noblest minds of the day, and to give them a clear and permanent insight into a supra-animal reality, it seems that far too much theology and ritual were allowed to remain. The time had come for any technique which was to be decisive to be purely psychological. The effect upon character is just what we should expect from a process which is half superstition and illusion and half psychological and objective. The best men, realizing best the vital need of an enlargement of consciousness, the clear revelation of an objective sanction for supra-animal conduct, undergoing the initiation, gained sufficient assurance for them to carry on. They do not and cannot gain enough to raise them to a new way of living. Plato himself, the most influential of all the initiates, could only advocate as his final social contribution (The Laws) a secretive and cruel tyranny, believing in an ideal end but using, able only to use, violence to gain its end. Aristotle—a kinder, more stable, less urgent nature—can only recommend, as an alternative to a society run by Inquisitors, one framed on the Mean, a cautious balance of virtues, that mutual self-interest which is the vain hope, but the only possible hope for all utilitarians. Aristotle, according to Synesius, owned that "Initiates do not learn anything but receive impressions."

Does this mean merely that they witnessed an impressive ritual, or passion play, or does it mean that they underwent some effective and lasting hypnotic suggestion?

Meanwhile actual contemporary civilization steadily declined. Plutarch, one of the last of the eminent Greek authors who accepted the mystery technique, as the one hope of mankind, spent his life during the early Roman Empire, carrying out the duties of a minor provincial official, his leisure taken up with the praise of the heroic past. His acute mind realizes that as outsiders question and inquire about the mysteries, the holy ritual will be dismissed as merely agricultural rites—fertility religions which a romantic movement is reviving out of archæological interest. He acknowledges that initiates can produce no demonstration or proof of the beliefs they acquired. But he maintains that the mysteries are not farmer's magic, that the initiate does "gaze on unspeakable beauty." This indeed may have been true. It is, however, even more certain that he could only bring back a private comfort. He could neither make outsiders able to endure tyranny without damage nor, himself put in power, could he carry on a rule of justice without employing violence. The only other sanction, the only non-violent, non-self-defeating sanction, given by open vision which sees through Time and so can demonstrate a fulcrum beneath Time's flux, he did not secure.

Here, then, the Greek exploration ends. Henceforward religions of increasing authoritarianism and other worldliness will dominate all Mediterranean thought.

CHAPTER IX

THE HISTORICAL EVIDENCE FOR THE EVOLUTION OF CONSCIOUSNESS

IV—Middle Ages

THE ACTUAL generation when this revolutionary retreat took place is difficult to date. Like most changes in mental attitudes and outlooks, it certainly occurred long before it was recognized by those who underwent it, and considerably before the date given to it by party historians. The Christian Era was by all such supposed to mark, and was accepted as marking, that break with the past, when Hellenic thought ceased to be the inspiration of the intellect and Hebrew morality the ideal of righteousness. Such a line historians have long known to be arbitrary. It is both too comprehensive and also it is placed too late. Reefs of scientific physical research ran out into a rising sea of intellectual indifference, as at Alexandria where in the first century A.D., Hero with his ælopile invents the first turbine, and, as we shall see, at Damascus physical and psychological experimentation flourished certainly until the Arab invasion and probably after, perhaps only extinguished by the Crusades.

There were, therefore, many exceptional surviving advances going on into the Dark Ages. Nevertheless, a retreat did set in. Moreover, that retreat was definitely begun long before the beginning of the Christian Era. There was a changeover. The aim, however unscientific, however unconscious, had been to take part in evolution by continuing the enlargement of consciousness and so to come to a clear, consistent view of Reality. The aim which succeeded to this was first to accept the distorted and partial impression of Reality which the untrained, undeveloped senses gave as being complete Reality itself, and then, finding that such a picture was intractable to man's wishes and did not make sense, to retreat in dismay and to hope for another world in which the environment would be more amenable and friendly. Some modern historians believe this change to have been already in the air in the time of Socrates. It seems, however, first unmistakably evident with the Stoics. Then the process is straight and clear. First Stoicism, the system of the Syrian, Zeno, a man who, we must remember, is a Semite not a Greek, will teach an iron resignation, and then Christianity will preach that the world is fallen and will only be restored by a cosmic cataclysm.

Stoicism will have as a technique of training nothing more than the simple contemplation of "Nature's perfect and unwavering Law"—facing facts with the pride of feeling that one has the courage to see things as they actually are. Here is no realization that it is our degree of consciousness which limits what we see, and, could it be extended, we should perceive another order of being, a further range of reality. Stoicism, therefore, could do even less than the Mystery Religions to reverse the degenerative change in civilization. The Stoics are strong enough for some six generations to man the ship of State, to keep it running (to maintain for lower types a social service which, as has been said above, will provide for their lower needs, for a certain degree of order and a "dole standard" of physical satisfaction), but not to steer it off the rocks and, when no one can any longer deny that these lie directly ahead, Stoicism can only council abdication and suicide.

Christianity begins by being *"charitic,"* a faith based upon the intense experience of affection which may be felt by those in a small group. Here is a very simple technique and proof: "We know that we have passed from death unto life because we love the brethren." Undoubtedly such a system works, if the numbers of the group be kept small and the constituent are simple: as undoubtedly it fails when applied to large numbers and highly self-conscious individuals. It works all the better if all society despises and persecutes the "little flock" and the flock itself is convinced that all society and the world itself are at any moment to be ended. No acute administrative or economic problems arise. The group is welded together by outer opposition. All internal problems are arrested: nothing need be thought out for "The Lord is coming." Undoubtedly such a system fails when applied to large numbers of sophisticated individuals, who join it when to belong to a Secret Society is more of an adventure than a scandal, and when the ardent belief that the world is about to explode has been damped by generations of disappointment. A numerous and elaborate society must have a consciously defined and clearly analysed technique, based on an evident empirical psychology. The psychology of any epoch must be at the same stage of advance as its economics and physics, if a serious regression is to be avoided. Our perennial tragedy has been that our working psychology is always a whole epoch behind our physico-economic state.

It was therefore inevitable that Christianity should change, as a consequence of two generations of success, from charity to mortification. The discovery that simple love was not enough and that the world was not coming to an end might have stirred the Christians to careful and earnest exploration and experimentation. Civilization was bankrupt and in liquidation, but mankind remained, and was about to call on Christianity to be its "competent receiver." The Church should, by devoted research, have worked its empirical psychology into a science. Unfortunately, the reverse happened. An iron authoritarianism ended all hope of further psychological discovery. The Church rejected Origen's philosophy and Marcion's advice and tied itself to the Jewish anthropomorphic cosmology. Nor is the further contribution which Christianity received from Neo-Platonism—the last addition to its thought—any aid toward it becoming experimental and scientific. Neo-Platonism has an eloquent sublimity with which to persuade the individual to seek enlightenment but that system not only rejects the world as an evil dream, it also fails to give with any precision technical advice as to how man may precisely train his consciousness so that he may have open vision of reality. Plotinus (A.D. 204-270), just as much as the Mystery Religions' proficients, only uses the ancient machinery of a decayed theology and (here he is behind the Mysteries) he makes no attempt to give people initiation. His whole system, as a practical working technique, is no more than imaginative rhetoric based on an outworn symbolism, subtle (but it must be confessed often unconvincing) argument, meditation on noble themes and then the hope that perhaps once in a lifetime may come spontaneously a sudden blinding flash of insight, so sudden and so brief that after it the man can only say it was utterly incomprehensible. Porphyry, his chief disciple and editor, declares that his master only attained ecstasy (Illumination) four times, to his knowledge,[1] while he himself only once and not till his sixty-eighth year.

Though, however, conscious experimentation was closed, the mind of western man continued to wrestle blindly with the problem of his inner nature. We know that in the Egyptian Thebaid—probably influenced by the Therapeutæ but taking on a more desperate complexion as the monk, in contrast to the Therapeutæ, was often a young and ignorant man of high sexuality and low intellectual tolerance—Christianity took on a strange and significant form. There the Eremites attempted for some four centuries to force the trap-door of consciousness by mortification. They seem from their writings to have had little notion of what they were doing or of how it should be done. Their advices are keyed by the principle that a "diseased body means a healthy soul," and they used as mantric aids (those repeated phrases whereby most mystics shift the focus of consciousness from "comparative" meditation to self-transcending contemplation) only the Hebrew scriptures—the psalms especially, those songs which, we have seen, are, because of their passionate nature, peculiarly unsuited as mantras. Nevertheless, one oriental scholar of unusual information believes that they may have known and practised Yoga methods. The fact that their predecessors, the Hermetics, were as we have seen, undoubtedly in touch with India would lend some support to this opinion. Another reason to suppose that their intense mortifications were something more than masochism or merit-making, was that their practices passed half-way round the ancient world. There can grow from Fakirism a certain degree of Yogism, indeed more easily than can Yogism grow from the later procedure, petitionary prayer, which is the procedure prompted by anthropomorphism. Though the Church was never at ease about such extremists, their practices spread very widely on the fringe of its authority. In the sixth century Ireland gladly accepted twelve monks from Egypt.

1. *Life of Plotinus,* Cap. 23.

Celtic Christianity becomes intensely mortificatory. The rule of St. Brigid, "the Mary of the Celts," is almost a flagellant system. This, then, is one more return to Fakirism, the crude attempt to enlarge consciousness by pain.

Beside this effort to force consciousness by complete sex repression, severe abstinence, constant discomfort and frequent pain, there was also spreading in the northwest some more rational psychical exploration. To indicate the origins of this is even more precarious. As Dr. Inge has remarked, "We are baffled both by the loss of documents and by the extreme difficulty of tracing the pedigree of religious ideas and customs."[2] We do know, however, that there was a line of active and acute thinkers in the Levant who hoped to be able to use Christianity as the channel in which all the psychical knowledge of the past might be brought together, and, so combined, give rise to a constantly growing psychology of the future. Such a pioneer is Origen. He attempts to rid psychical thought of dependence upon and confinement within "Contingent facts of History," its rigid imprisonment in dogmas about historical or pseudo-historical events. His method was that already used by Philo and his Therapeutæ in their attempt to develop Judaism: to maintain that all the dogmas and indeed scriptures could be accepted by allegorizing every incident. He speaks with unguarded pity for those who accept such writings at their face value, and take such stories as objectively true. The common people had better do this. That is the "milk for babes." The wise, those who have the Gnostic insight, read under the lines and the obvious meaning. "His gnosis neutralizes all that is historical, if not always as to its actuality, at least absolutely in respect of its value. Historical Christianity is stripped off as a mere husk."

The Church was long at a loss what to make of this master, at once so useful against enemies, so dangerous to subjects. At last when, safe in the saddle, revolt of subjects is all that is to be feared, under Justinian, Origen's writings are condemned. His less intellectual, more vague contemporary Clement of Alexandria (A.D. 150-210?) also attempts this method of preserving outward conformity

2. *Christian Mysticism*, p. 350.

and retaining for the educated, inward freedom whereby "the exegete reads his own ideas into any passage he chooses" He, too, was at first deeply respected, much used and finally degraded, losing a saintship, long conferred upon him, centuries after his death by authoritative act of a late Pope (Benedict XIV, 1740-1758) as the Church finally contracted.

The third of these remarkable bridge planners if not bridge builders, succeeds in being acceptable if not efficacious. He however took the precaution of assuming an unknown ancient worthy's name. "Dionysius the Areopagite" claims to be Paul of Tarsus's rare convert in Athens. He was, it seems clear, a Syrian monk who popularized the work of his master whom he also disguises under the vague name Hierotheus, but who was probably called Stephen bar Sudaili and who lived about A.D. 450. Here we have the same attempt to rid the intelligence from the censorship of the superstitious, and the exploring mind from dependence on authority and limitation through canonized tradition.

The curious success of this venture, that this daring speculator, this thinker wholly concerned with attaining direct experience of reality, about which nothing must be predicated, not even that It is Good, not even that It Exists, should actually be canonized, is not explained merely through his forging a Biblical name. Origen failed with a more flexible Church and Clement awoke a lukewarm reception, because they did not give sufficient sop to the ecclesiastical Cerberus. "Dionysius" took care to do this, coating his psychological speculation so thickly with "proofs" that the earthly hierarchy was a manifestation of the divine, that the Church authorities bolted the tempting morsel—the little book which like that of the Apocalypse was sweet in their mouth but bitter in their belly. The Council of Constantinople in A.D. 533 could not resist quoting him as an apostolic confirmer of ecclesiastical claims and once quoted with approval by a Council, the vast machinery of the Church can hardly go into reverse.

Yet this method of introducing an ingeniously concealed "relieving arch" behind the heavy facade of petrified dogmatics, so that under its cryptic protection the thinker might explore, in the body of the Church, but free of its confinement, this method was not efficacious—at least where the Church could really rule. The solution of allegorizing away dogmas postpones, it does not solve the problem, the issue between a blind faith in authority and a working experimental psychology. It may be compared with that modem dangerous tendency in aging democracies: the refusal to repeal an anachronistic Law because of the disturbance and difficulty of so doing, and the getting round the problem by letting the Law become administratively a dead letter. This is highly dangerous because freedom becomes dependent not on free knowledge of clear rights but on an unwritten agreement by a few civil servants that the Law should not be worked. At any moment these authorities or an ignorant public opinion may revoke what is only an unauthorized licence. This happened in the Church. Origen was first condemned by a venal archbishop, yielding to ignorant fanatical monks.

Within the Church thinkers will repeatedly try to avail themselves of the rights which it was thought Dionysius had won for them. The greatest of such explorers is probably Eckhart, who in the XIII century attempts to make a free religion of experimental experience. He is condemned. Only on the borders of the Church, out in the far northwest, where Rome, and indeed all legalistic centralizing authority could be disputed, did the work of Stephen bar Sudaili, the Syrian monk in touch with Indian ideas, begin to work toward further efficacious exploration. The thought of the speculative theologian Erigena (ninth century), working on the pseudo Dionysius, the Neo-Platonism which had so entered Christianity, and the oriental conceptions of transcended personality, relieved Celtic psychologists of any dependence on anthropomorphism. Not only is God no person: nothing can be said of him. He is something which is known because the soul finds itself drawn on into darkness. He is the extrapolated goal of a migratory instinct of the spirit. Such thought left the experimenter free to study the soul (psychology) rather than to study the aim (theology), about which, until the soul is further developed, it can only speak nonsense. The organized Church could not approve of this. Erigena himself is condemned. The experiments, however, continue and as one of the greatest students of this difficult phase of history, the Abbé Brémond, has decided, it was a small current entering North England from Ireland and thence passing into France which gave rise to the mystical movements in early modern Catholicism.

By the time we have documents written by such practitioners we can both trace that movement coming south and also we can see not less clearly how, as it comes south and emerges into the open, it loses increasingly its exploratory character, ceases to be an experimental and practical psychology and becomes orthodox, theological, anthropomorphic. Richard Rolle, the early fourteenth-century Yorkshireman, is entirely interested in the attaining of certain intense ecstatic states, in the time which such dilations may take, the methods to be used, and the psycho-physical conditions which are concomitants. The next known authority, the author of *The Cloud of Unknowing,* would seem to have written in East Anglia, found it apparently safer to be anonymous, warns his readers strongly against lending or even handing the book to any third party, is less interested in the psycho-physical conditions and makes more assertions of his orthodoxy. Nevertheless, he is still empirical, only interested in the dilation of consciousness and contemptuous of anthropomorphism. The Church, however, was becoming increasingly suspicious of such empiricists, and we can appreciate the growing necessity for the caution he enjoined when we note that one of his typical exercises for surmounting the ego and attaining wider consciousness, "I have nothing: I know nothing," is glossed in a later edition, "Save Jesu." At once empiricism is abandoned for an *a priori* conception of personal salvation. The third of such writers, Hilton of *The Scale of Perfection,* is from Northampton in the fifteenth century, acknowledges that of the higher, rarer states of consciousness he has no personal experience, is far more exacting in making preliminary claims for strict orthodoxy and throughout his prolix book tends to introduce a modified anthropomorphism wherever he can. We can also trace this development a step further, for in the successive editions of this medieval best-seller—it was among the first books to be printed—the name Jesus is increasingly interpolated. The slightest suspicion that there is being advocated any method other than that of grace through petition must be removed.

We can say then that we have slight traces throughout the Middle Ages (prior to the swift triumph of the analytic method which characterized the Modern Age) of uninstructed efforts to enlarge consciousness. Further, these efforts may be divided into two classes, the fiercely mortificatory, pain and misery being held to be forces which can loose the soul, and what may be called the vigilant or "stalking" method, whereby the body and the mind are cautiously and deftly soothed. This is the reverse of the mortificatory. It atrophies, it does not amputate. It wins by skill and subterfuge what the Fakir would seize by violence. We have seen the "Fakir" Elijah is followed by the "Yogin" Elisha who employs a specific technique. Such a method always follows the Fakir's. That was part of Buddha's success. He was more scientific, more psychological than the mortificators. These later teachers of the Way of Right Knowledge have discovered Baudouin's Law of Reversed Effort,[3] that the harder we wish with the surface will, the more firmly the subconscious will resists and acts reversely. They have learned that the valve of the subconscious is more often jammed then pried open if it is blasted with pain. It seems strange to us, who imagine that all medieval mystics must have been mortificators, to discover that there was a school which had found a path leading to more accurate and more extensive results. There can, however, be no mistaking the fact that the author of *The Cloud of Unknowing* knew this "more excellent way," for he specifically warns his correspondent against tormenting the body and even begs him to look after it carefully and nurse it wisely out of any sickness. While it is ill, any enlightenment which the soul may have, is suspect. Illness comes under the same disqualification as an aid to enlarged consciousness as do self-induced pains and drugs. Possibly they help; it is, however, very difficult so to control them that they may rouse awareness but not then confine the attention so roused to themselves.

There were, then, being made in Europe until the beginning of the Modern Age, two persistent, but uninstructed efforts to continue evolution through dilation of consciousness. The one was more

3. See Chapter X, pp. '75, 187, and Chapter XI, pp. 19o-191.

admired by the people and approved by authority, that was the Fakiristic. The other was nearer to Yoga, and therefore unimpressive to the masses and suspicious to the Church. This must always be so, because common men are more impressed by violence, if it be nothing more than self-violence, than by accurate action, and authority finds it easier to deal with a man whose primal motive is to suffer, than with one who is fundamentally inspired by a desire to understand.

After the Renaissance, the rise of individualism and of analysis, the spontaneous Yogic effort declines and maybe vanishes. In the mortificatory branch, however, there is some development, a development which it may be of interest to note, continues its path southward. France, Italy and Spain produce contemplatives, men, who by severe discipline, aim specifically at an experience which they state clearly that their present state of consciousness cannot apprehend. The mind must be cleared of images, that is their peremptory demand. The Counter-Reformation successfully enforced a contradictory system. The soul, as much as the Church, must be crammed with images. The generalship of Ignatius recovered for Rome the greater part of the territories it had lost. It is not surprising that he won also on the frontiers of the mind, there ordering retreat. Even in Spain[4] there is clear evidence that much greater freedom of the spirit existed before than after the Reformation. After, the mystics in the Roman Church were permitted to develop to a point, because their record of sanctity was useful to Authority through giving the Church spiritual prestige. The established Churches use the copyright of the mystics to claim that they themselves are a source of spirituality and the later mystics are too intellectually unaware to check this dangerous "piracy." Once the Reformation is in retreat the Church crushes these dangerous servants, only sanctioned during a crisis when concessions had to be made.

John of the Cross dies in penitentiary confinement. Theresa owns that she dreads the men of the Order of Jesus, and her work only reaches the world because it is heavily anthropomorphized. Fénelon

4. See Inge's *Christian Mysticism,* p.214.

falls under the gravest suspicion. Molinos is condemned to solitary confinement for life, and so dies. There is nothing unexpected in this shackling of those last experimenters. The battle between psychologists and theologian, explorer and authoritarian, the man who aimed at the evolution of consciousness and the man whose one aim was to preserve his self-conscious ego in another three-dimensional world, through forgiveness of an offended personal God—that battle had been lost by the fifteenth century if not before. From that time the Church had increasingly approved physical representations and corporeal devotions. The Five Wounds as a cult is a popular precursor of the Sacred Heart, and before that there is a specific Jesus Worship. On the other hand, the explorers were uninformed and very unwilling to break away from the anthropomorphic Christian tradition.

As we have seen, the tradition of experimental religion, the pursuit of the dilation and evolution of consciousness did pass down from Ireland to England and was still going south at the end of the sixteenth century. The Abbé Brémond in the second volume of his *Histoire Littéraire du Sentiment Religieux en France* (pp. 152-168) has shown that a strong and effective mystical force entered France from England about 1570, when William Filch, a man from Canfield in Essex (and so-called by the French, Benoit de Canfield), had himself priested at Douai, became the spiritual Director of a number of brilliant and influential men, such as "l'Éminence Gris," Richelieu's "demon," Pére Joseph, and wrote for them a booklet of mystical theology called *Régle de Perfection.* Though he attempted to teach a double theology, reserving the third part of his book, wherein anthropomorphism and indeed theology are dropped for pure psychology, he has been charged with Quietism and one of the renderings of his text was in 1689 put on the Index.

Probably the charge is just. For Quietism was an inevitable development, an unavoidable stalemate when there was no adequate psychology to develop thought beyond individualism, and thus to achieve a praxis to develop consciousness beyond the ego, The Quietists, from bar Sudaili to Molinos, know that they make contact with something far more real than anything that is apprehended in this temporal experience. There is, they know, beyond space and time, but immediately and always present, a state of being, an enlarged consciousness, beside which our present normal experience is a fevered dream, the experience of a person mentally sick. But because neither their theology nor their psychology has any terms for this, they can only describe not only the goal but the activity toward it, by negatives, as complete inaction. Their phraseology could not be accurate. It was all sprung from anthropomorphic concepts. Had they defined their aim and process more exactly they would only have been the more easily and quickly condemned. For they had, and could have, no place in an authoritarian, traditional Church, tied to historical dogmas. Had there been psychologists present, instead of jealous and timid onlookers, they would have understood that, though the phraseology was inaccurate and inadequate, some important experience had undoubtedly been stumbled on, should be investigated and might be defined. There were the facts to which the Quietists and their contemporaries bore witness: the wonder and delight they experienced; the joy, the simplicity, the influence and the innocency of their lives (for the charges of immorality brought by the Inquisition against Molinos cannot be given weight considering what the Pope himself had thought of him, watching him while the Quietist actually lived, by papal invitation, in the Vatican, what thousands also confirmed, and, considering both what the Inquisition took as evidence, the legal procedure of the Holy Office, and how, once an error was made in theology, physical sin was assumed in the heretic's life). But fear and resentment were ruling the world. Even the Saints in the opposite camps, Protestant and Catholic, feared each other and sought safety behind ecclesiastical defences.

Hence, even if the Church had not molested the Quietists, it is doubtful whether they could really have aided the further evolution of man. They themselves were probably wholly unaware of and unprepared for the sacrifices, "purgations," "aridities" and "dark nights" which further advance, had it been permitted them, would have entailed. There are worse things than confinement when a man wants to fly but has as yet grown no feathers on his wings. In the middle of the next century, William Law, having given up rank and position, influence, office and salary, retires into the depths of the

English countryside and there with the aid (or handicap) of the books of Jakob Boehme, attempts to continue this exploration. He cannot achieve it. He is aware, as is the migrating bird, of whither he must go but, not only can he not state the goal (no one can), he cannot define the means whereby the journey is made, save in terms which are already archaic. It is that statement of means in modern, accurate terms, which is possible. But it is only possible by those who realize that self-consciousness is a phrase in the evolution of the psyche and Time is a distortion of experience caused by that strangulation of consciousness which we call the ego. The very idea of such things was more alien to the Protestant than to the Catholic, more painful to the devotionalist (the bhakti) than to the traditionalist. Hardly any of the Quietists had reached beyond a passionate private devotionalism fixated on a person, however little they might define that person.[5] Hence, they discovered no clear technique or firm knowledge before Authority came down upon them, offering them liquidation or a whole-hearted return to the Party Line.

The Church, however, had won a fatal victory. The majority of western mankind, compelled to choose between a personalism which was for all practical purposes, anthropomorphic, both in its object and in its technique, and analytic mechanism, chose the latter. At least it gave them more efficient means, left them free, was easily comprehensible and more true than anthropomorphism.

It may be questioned why in this outline of medieval attempts to dilate consciousness—cursory as it has had to be—there has been no mention of Jan van Ruysbroek, "Doctor Ecstaticus" (1293-1381). Certainly here was a natural psychological genius of the first rank and one who described his extensive explorations on the frontiers of consciousness with copiousness. Certainly, too, he understood to an unusual degree what was the psychological factor involved in his experiences. Taking his nomenclature from his immediate master, Richard of St. Victor, he not only uses the illuminating phrase *"mentis dilatatio"*—the expansion or dilation of consciousness—but, what is of peculiar significance, because it shows that Ruysbroek understood

5. e.g. John of the Cross' hymns.

the supreme importance of that fact (that the focus of consciousness can and must be shifted and expanded) he makes this *mentis dilatatio* the cardinal condition and experience. Richard of St. Victor had named as two successive states this *mentis dilatatio* and then, above it, *mentis sublevatio,* when the consciousness experiences an uplifted, transported condition. Ruysbroek daringly reverses his master's order. To him a shift of consciousness, so that a new focus of awareness is attained, is more important, more significant and of more finality than a state of disturbance, however ecstatic. Nevertheless, in spite of his magnificent endowment of natural gift and his artistic descriptive power, he remains an artist (and one, also, strictly confined to traditional forms). He is not in any way a scientist, a true introspective psychologist, and so adds nothing to the science of the evolution of consciousness. He is like a man who with great energy has won to the Pole and with a vivid descriptive style tells us of the Arctic Night, the Midnight Sun and the strange splendours of the Aurora Borealis. But, having no scientific instruments, he can give us no accurate bearings, no charts nor maps, and only a rudimentary description of the equipment and diet needed for the voyage. His tradition forbade such exactitude. Like another explorer, Columbus, he was compelled to call his New World by old inaccurate names, and with even more unfortunate results than befell Columbus. For though Columbus did not discover a new route to the Indies and the shores he landed on were not Asia's, as he had hoped and maintained, it was clear to all that he had found something, he had landed on a new continent. With Ruysbroek and the orthodox school of mystics, because they called the new world of consciousness which they touched, by old conventional names, those who realised that the names certainly did not fit the reported facts, then denied that the mystics had discovered anything at all—except perhaps a method whereby through balked sex to bring on autosuggestive hallucinations. Ruysbroek, then, remains a great artist of the Upper Mind and Growing Edge of Consciousness, and not a scientist, because he is tied to tradition which is frankly afraid of experiment and of the law of parsimony. Certain experiences cannot, must not, be taken in themselves. They must be assumed to be given by a Deity whose Nature the theologians had logically induced from examples found in a collection of books of vastly varying moral tone but all declared to be inspired and infallible on grounds which no mystic, left to himself, would or could have allowed. Once this assumption is made, experiment becomes highly temerarious, if not blasphemous, and too close attention to method is strongly disapproved. In such a climate a scientific technique will never grow. Ruysbroek's mental ancestry runs back through Richard of St. Victor, through St. Bernard, pillar of orthodoxy to the dogmatic Augustine, who owns that his only notion of psychological method—Introversion—came from reading the NeoPlatonists—and how inadequate their methods were we have seen. When to such a rudimentary procedure was added the handicap of a strict, rigid theology derived largely from Hebrew anthropomorphism, it is clear that however great the natural genius of such a man as Ruysbroek, he could not add to psychological method for the evolution of consciousness.

PART III

THE DISCOVERY OF TECHNIQUES

CHAPTER X

INDICATIONS OF PAST TECHNIQUES

SUCH then, in briefest outline—and one often broken, through lack of record—is the history of the enlargement of consciousness, as detectable in the West. Scanty though the references be, and difficult as are some of them to interpret, we can maintain that man's specific evolution, the development of consciousness, has left sufficient traces of the path it took. We can conclude that there is an urge within him to dilate his awareness and to alter the aperture of his apprehension. We can show that those peculiar symptoms, human pain and human lust (because of their significant difference from ani-mal pain and lust), indicate a balked reservoir of vital force, which, when permitted to energize higher faculty, relieves the body from anguish and conflict and gives the mind direct apprehension of a supra-animal, supra-temporal world. We can first detect the widening of consciousness at the psycho-physical level. We can perceive that, under the stress of constantly widening interest, man has found that his eye developed an extending colour sense. Hyperæsthesia, especially of that sense which predominated because of its superior range, has evolved under the pressure of constantly expanding attention. Next, we notice appearing, alongside of this supra-sensory perception of the body-mind, a further development: the emergence of a purely psychical faculty—extra-sensory perception.

These two developments are undoubtedly associated. It was first necessary to detach man from his physical interests which were engrossed by the short-ranged senses of immediate satisfaction—touch, taste and smell. Sight is, therefore, developed. So man becomes increasingly aware of perspectives which dwarf his actions and drive him to alert passivity. At the beginning of civilization he becomes acutely interested in the stars—phenomena to which all other animals are indifferent, and of which all man's other senses are ignorant. Siderial vision gives him a profound sense of larger significance lying outside the world he can handle. So he cannot help imagining that it is in the unattainable sky that his destiny is settled.

Once, however, sight has delivered him from exclusive attention to phenomena to which he can make a bodily reaction, sight itself must cease to be the growing edge of his enlarging consciousness. Sight is a stepping-stone leading man from the body to the mind. It is a hyphen between the purely physical and the purely psychical. Man has, therefore, to make another radical mental departure, a mutation of attention, even greater than that which raised him from engrossed absorption in physical contact to detached and patient wonder. He has to break with his past more completely than when he rose up from animal to man. He has to become aware of the illusion of Time, a more radical and more difficult revolution in thought than any of his revolutionary ideas about space.

There are, then, three great stages in man's evolution; the appetitive animal, the intelligent human (Nietzsche's "transitional creature"), and the being who once more knows itself to be a part of the Whole, but this time is clearly aware that that Whole is a comprehensive consciousness.

Sight is the specific sense of the second stage, that of the intelligent human, as smell is the specific mammal animal sense. After having done its work of freeing man from his animal matrix, it should itself yield to a still higher form of apprehension. It must, if human evolution is to continue. Unfortunately, all forces appear to gather excessive momentum. Having carried man out of his animal groove, sight begins to dig for man another rut even harder for him to surmount. So greatly enlarged are his powers (his means, if not his satisfactions), and so vast now appears to be his outlook, that man becomes intellectually certain that henceforward he sees the whole frame and limits of reality. Seeing is believing. Hence any apprehensions which cannot be tested by sight—by "pointer readings"—are treated as illusory. Sight, therefore, by its exclusive dominance, after it had freed man from his animal adhesion to, and kinship with, the world, carried him too far. He became convinced that he was completely separate from the world, and so that what he saw was the actual and only possible reality. From this he assumed that the universe was an inhuman machine, that he was free to behave as he liked, for there was no intelligent being anywhere save himself, and that the only way of affecting reality was by physical effort—an effort which, though it could destroy his own handiwork and his civilization, altered only insignificantly the apparent universe.

This whole line of thought was, in short, a degenerative specialization and has led in consequence, in our age, to a moral breakdown. Man has become individually a creature filled with acute conflict in himself and so psychotic; has become socially a species which has not enough power to alter the Nature it apprehends, but has enough power to destroy itself, and sufficient blindness to sink into intraspatial struggle, and so internationally anarchic; has become intellectually contradictory, believing in two mutually exclusive experiences—values and reality—and so has abandoned reason.

If, then, man is to continue his evolutionary advance, he can only do so if he "springs again from below the graft." That is why we have had to search in the period prior to the rise of what we may call sight's false extrapolation, for a new start, for a new advance in which we may recover our contact with reality, a contact which we have largely lost because, without knowing, we passed from the precepts of sight to the concepts of seeing, from what we actually see to what we think we see. This idea, that we distort what we see by our unconscious assumptions about it, has been known for centuries to the artist. As we have seen, analysis ruined Neolithic art, depriving man of his former power to see objects as an organic whole. In our own age the great struggle of the new art, which dawns with Constable, is to see as he said: "There are no boundaries (lines); there are only series of coloured blends and contrasts, every object throwing colour or receiving colour from the rest."

The practical man has, however, dismissed all this as simply the nonsense of fanciful artists who can't draw. With the advance of physics, practical men can no longer dismiss this problem of seeing—what we see and how much we impose, how much we spontaneously, unconsciously fancy. Einstein himself has pointed out that the reason for the difficulty of understanding the Principle of Relativity is in the main due to our unconscious assumptions that there are such things as mass and time, when in fact there are only objects that are heavy and experiences which last long or are brief. Professor Dingle has pointed out how much easier it is to teach boys, who have never been taught to take the Euclidian assumptions as true, to understand Relativity, because they more clearly see things as they are and not as geometrists imagined them to be. When we realize that the stars, as an actual experience, are minute to the point of insignificance, and that the whole mechanistic picture of space and super suns, is a deduction and not an experience; that the notion of the zero cold of space might just as well be imagined as a furnace heat,* for no feeling has ever sensed that condition and never will, we begin to realize how much our eyes have run away with us, or it might perhaps be said, away from us. We must then come back to direct and whole experience, prior to the over-statements of the analytic, mechanistic phase, for evidence of those buds of wider and more advanced (if less defined) faculty from which further enlargement of consciousness may spring.

These earlier, pre-analytic men, although lacking our present power of experimentation, had not made the mistake of breaking up the one reality into two unconnected parts: they had not fallen into the error of assuming that the apparent world of common sense was final reality, and that the consciousness of man had no part in making that appearance. They were, therefore, able to give themselves, however inexpertly, to the task of understanding consciousness and to attempting to alter, improve and clarify it, as a first essential step to understanding, and, if need there be, to altering the phenomenal world.

In the last four chapters a very slight review has been given of such experimentation. The record of this psychological exploration has been taken from the histories of Egypt, Palestine and Greece, partly because from these three sources are derived our own theories of human consciousness and how it may be enlarged, but mainly because these particular histories can be dated. We can therefore, trace in succession the stages of man's specific psychic evolution and, from such a series and trend, we may be able to estimate in

*See Sir Arthur Eddington's *Nature of the Physical World.*

what direction and by the development of what means, our essential progress may be resumed. This historical sketch has been brought down to the beginning of the Modern Age. There it must halt, for as said above, with the use of physical analysis, with the assumption that the outer world is directly "given," and is a clear, hard, objective reality, the appearance of which owes nothing to man's mind, there ceases the exploration of consciousness and the attempt to expand it.

To-day, however, we must certainly ask: Is this the end? We cannot doubt that man's true aim and Life's intention for him is not (as the Modern Age has assumed) power over his apparent environment. That environment is no firm, tangible surrounding on which he may work. The very dominance of the sense of sight, a triumph on which his physical science wholly depends, defeats all hope of physical mastery of the world so revealed. As we have seen, the universe which his eye discloses is an environment immensely beyond any hope of his control. Great as has been the advance of his powers, in comparison with the range of his sight, these powers are insignificant. Indeed, we may say that in comparison with the world which the animal sees, and its power to alter that world, man has become, proportionately, more helpless. In short, the sense which has grown fastest and has become completely predominant is the sense which stultifies his physical power. On the other hand, while the progress of his physical senses shows him increasingly a world which his physical powers can less and less hope to control, these physical powers increase rapidly enough to be the gravest danger to himself. They will never permit him to control his environment. They do allow him to destroy his civilization and his humanity.

These facts are now generally recognized, and have caused men to abandon the old uncritical belief in progress, a belief due to the eye's unanalysed ascendancy. If physical sight is seen as a stage toward, but not the final apprehension of, Reality, then we see Life has an aim. What that aim may be is indicated by other evidence, little recognized but quite definite. We have seen that, beside the expansion of his physical vision, man has experienced throughout his history outcrops of a profound insight, extra-sensory perception, direct apprehension of a supra-animal reality through a supra-physical, supra-temporal awareness. We have also seen that those men who were most profoundly affected, have striven to hold these moments, to become native of that larger world. They realized that without such open vision the people perish, civilization is doomed, man must become extinct.

It, therefore, does not seem possible to doubt that it is life's intention that we should go in the direction of enlarging consciousness. Nearly every observer is convinced that all other lines of advance are closed. Without a new psychological mutation our race appears to be arrested and doomed by that very excess of unbalanced growth which has marked the doom and caused the extinction of all the other earlier dominant animal species. Nearly every researcher into human faculty is also persuaded that the deliberate enlargement of consciousness is possible, and that a natural tendency in that direction can and must be cultivated until what is sporadic and involuntary becomes continuous and intentional.

Western man must then resume the task which was interrupted by the triumph of the analytic method and the rise of mechanism. After three centuries, during which the most active communities of mankind have worked with a profound and increasingly deadly misapprehension of reality, we must go back picking up once again the true line of advance. This is, however, no blind reaction nor need we regard as wasted all the devoted labour of physical, analytic research. It is, perhaps, necessary here to repeat that what is now being advocated is not a blind return to the past but, on the contrary, a taking up again of themes which had been prematurely neglected and developing these with the same intense precision as physical science uses to-day in its researches. Indeed, one of the specific characteristics of contemporary science is precisely this procedure—the searching in the past of each science to see whether, at every decisive turn in the road of exploration, the road which was *not* taken, the hypothesis which failed to find majority support, did not contain certain observations, then neglected as futile but now full of suggestiveness. Lorentz, the Dutch mathematician to whose work in pure mathematics Einstein has said he was indebted when working on his Principles of Relativity, destroyed most of his original calculations, keeping only the conclusions. This, Einstein maintains, is a loss. In the steps which Lorentz took might have been found many fruitful lines of departure, indicating other conclusions as original as those at which he arrived. We may then say that, excessive and perilous as has been our specialization in mechanical analysis, nevertheless, the patient experimental accuracy and the critical detachment which its technicians advocated and often practiced, is a vital contribution to the further development of consciousness, to the study, practice and development of extra-sensory perception.

The great analytic method taught mankind the value of detached experiment and careful comparative record. Those in the past who worked at the task of expanding consciousness, worked as artists, not as true experimenters. They worked by tradition and intuition, holding fast to everything inherited and fearing to look too closely into their own methods lest they suffer by being understood. Indeed, we may hope that with the method of careful experiment, directed to the real problem which confronts mankind (the nature and limits of consciousness), we may make advances in the direction of ends, in the direction of the further growth of ourselves, even more sure and far more valuable than those hitherto made in the direction of means, in the direction of increasing supplies so that our arrested selves may be content with their stultified, frustrated state.

This is but to say, in terms of volitional discipline, and of deliberate mental technique, that as consciousness evolves we must become increasingly aware of that evolution and must increasingly co-operate with it, if it is not to fail. For, as far as we know, all that was required for super-sensory development—for the specifically human extension of sight and for the evolution of hyperæsthesia—seems to have been freedom to attend to the margin and growing edge of normal interest and curiosity; a certain sum of unstaunched vitality. No conscious progress in discipline was required. For extra-sensory perception it would appear to be necessary to throw out of gear normal sensory perception.

Moreover, as the further normal sensory perception has been specialized into complete dependence on sight, and that dependence on sight has led to the further mental restriction of mechanism, the more drastic has to be the effort before the mind can relinquish this extension and extrapolation. So the earlier, the pre-analytic type of man can dilate consciousness as the Canaanitish prophet, Balaam, is said to have done, "with his eyes open." Sight does not, at that level, paralysingly contradict extra-sensory apprehension, because to such a seer the findings of sight do not demonstrate mechanism and so do not prove to him that his vision must be nonsense. The "middle distance" type of man, the religious type who is trying to defend religion against mechanism by anthropomorphism, must needs counteract sight; so, first he shuts his eyes and calls in the aid of his ear by music, and later (a fatal retreat) he looks at an anthropomorphic image. The modern type of man, wholly assured of the truth of mechanism, can only have any apprehension of wider reality if he can have it at all—in and by complete dissociation. Shut eyes and music cannot help shift the focus of consciousness, already hypnotically fixated, by the massive social suggestion of his time, on the mechanistic picture.

We must alter everything; interest, attention, way of life, habits, views, our company, our livelihood if we are to have a chance of achieving this great step. There cannot, of course, be any certainty that we may emerge over this new threshold and enter on a type of consciousness superior to our present quality, as that is above an animal's. Because future advance must depend on conscious effort, on choice and intelligent striving, we cannot prophesy. We cannot say, even should humanity be convinced that here lies its true line of advance, that many will choose to take such a path. Nor, should men take such a line, is it possible to say whither that advance will lead them—at a pleasant pace through a series ever more purposive Utopias to a final perfection, or through a series of ever-intensifying crises to a sudden and radical deliverance, through a spreading social salvation or through so costly a strain that only a few will survive.

We have to envisage the possibility that though further progress is possible, it may no longer be possible as a double life. *Mens sana in corpore sano* may be an attempt at an arrested development and far from the purpose of evolution. Psycho-physical evolution, athleticism and insight may henceforward be an impossible partnership.

One day the amphibian which was to become a reptile had to abandon the border life of water and land, leave the shore and cast its lot definitely with the dry earth. It probably suffered much physical discomfort while completing this drastic adaptation, discomfort which the amphibians which remained arrested in the shallows have escaped. So, too, a time may have already come for man when, if he would advance and leave a frontier increasingly untenable, he must abandon half-measures, sever himself from the past and go on into a completely new condition of being. Even such a costly attempt as this may not purchase us the necessary advance. Do what we will, give what we can, we may nevertheless fail. Our knowledge of evolution does not reveal the goal or guarantee its attainment. What we can, however, say from our present knowledge of life's advance is that the continual striving after further awareness seems always to emerge on to a new level and expanse of being. If, then, we can continue so striving, we might rise. What is clear is that we cannot stay still. We must advance or suffer a catastrophic collapse. It is better to perish going forward.

Before, however, we attempt to see how the past empirical methods which we have noted might be applied as a science for the present, how from these hints and happy accidents we might make a psychology for the expansion of consciousness, we are compelled to add to the number of such rudimentary techniques as have already been referred to, one which is not represented in our Mediterranean tradition. This is taken from India, is called the Shakti form of Laya Yoga, and is the deliberate attempt to use sexual sensation and especially the orgasm to enlarge consciousness.[1] This Yoga is the converse of the Fakir technique. The Shakti uses sex to dilate consciousness as the Fakir uses pain. Recognizing that the orgasm is an uprush of primary energy which overcharges and so dissipates or divides consciousness, this Yoga attempts by foresight and intense continuous self-attention, to hold the mind in a state of detached interest during the paroxysm of excitation. Then the mind would,

1. For an outline of this Yoga, see Arthur Avalon's (Sir John Woodroffe) *Tantrik Texts: Kundalini* and *Shakta and Shakti* (Luzac, London). See also Vaijroli, p. lgg *et. seq.*, *Yoga*, Vol. 6, *Personal Hygiene*, by S. Yogendra.

instead of being swept under by the wave, be carried on its crest far above the high-water mark of normal consciousness.

It is obvious that the Fakir, with his extreme mortifications, attempts in a way closely analogous so to employ the pain paroxysm. Spastic agony, as we have seen, is, however, not sufficiently under control, and owing to the immense fund of evolutionary energy still pent within us, if this energy is unskilfully detonated, it may escape from the mind's control. The sympathetic nervous system, if unduly irritated, can behave like an ill-engineered mine, where one charge being lit, sets off all the rest and everything is wrecked. Even if the pain paroxysm can be kept channelled and focused, the effort is highly exhausting. The results, therefore, are disappointing, because the dilation of consciousness so obtained is both brief and ill-defined, and the attempts cannot be sustained. What is required is a constant state, and one which will not only last but endure during times of reflection and critical self-examination.

The orgasm, on the other hand, is generally more restricted in its release of nerve energy than is violent pain, and this explosion is also under the control of ordinary consciousness. Hence we have the experimentation with the erogenous zones. What is, however, probably the central discovery of the Shakti Yoga are the delayed and the arrested orgasms. Here there is a deliberate attempt to take the tide at the flood and when it would break and turn, to hold the extreme moment. These Yogins claim that in so making the reflective mind mount and master the emotional life, at the moment of its intensest instability, expanded consciousness is experienced. They also assert that the experience remains as a permanent possession, freeing the experient from the false belief that normal consciousness knows reality in the world of common sense. Finally they maintain that here is a technique whereby through a method of coitus reservatus not only is the body strengthened by a mobilization of all the endocrine secretions, roused, released within the body and not squandered without it, but that here is a way whereby those who find the struggle for continence too distracting, may achieve all the psycho-physical benefits without the struggle.

The subject is obscure and cannot be dismissed with a sneer about "chastity without tears." We know that many mortificatory agonies are necessary and even frustrating. However much we may admire the courage which will blow off its right hand in carrying out an inept chemical experiment, we cannot consider the experimenter enlightened, still less advise anyone to copy his procedure. Nevertheless, all these practices are predominantly physiological and aim mainly at achieving a body superlatively vital. In such a body, and to achieve such a body, there must be a highly trained mind, possessed of a considerably enlarged consciousness. But for what? For the maintenance of an indefinitely protracted physical fitness, a prodigious physical longevity (which some of these Yogins seem to have achieved), or for the future of humanity, for a new step in evolution? True, they used sex in a remarkable way, for a not unworthy purpose, but may not this way, which seems to make the path of the young aspirant easier, prove in the end a blind alley, so that he loses sight of the real goal?

It is hard not to conclude that in all these attempts to use the primal energy as it manifests itself through the sexual channel so that it raises consciousness to a higher level, we have a technique of extreme precariousness. Indeed, it may be said to be comparable with drug-taking for the same purpose. In both cases the individual consciousness is apt to be distracted and dissipated by the medium employed, so that, forgetting the purpose, the means become the end. Such methods have so shocked the conventional that we hardly ever hear of them save in accounts baited with disgust or blurred with gloating. In consequence they have been developed in great secrecy and under a sense of persecution and guilt. Nevertheless, even with our present unsatisfactory sources of information it is clear that in the Shakti form of Laya Yoga there was made a step in man's attempt to understand spastic sensation and to employ paroxysm to rise above Time, to enlarge, and not to diffuse or to arrest, consciousness.

Though such an attempt is an obvious step, we have seen we cannot place it in an evolutionary series. We can only presume from our knowledge of our own religious heredity, that experimentation in violent physical pleasure must have been contemporaneous with experimentation in violent physical pain. This would, therefore, date the research of this specific Yoga as beginning at the same time as the Fakiristic. If this line of experimentation was ever introduced into Europe, it must have been long after it had been fully developed, and maybe had reached frustration, in India. India could accept such exploration without disgust, or indeed, much excitement. The Puritan Revolution never swept India until the Mohammedan invasions. Buddhism advised strongly against all sexuality, but, teaching the overcoming of aversion as an essential discipline, was as incapable of disgust as of sniggering pleasure. Indeed, when the Larger Vehicle, Mahayana Buddhism, arises, that complete Union of Opposites, which is the end of all craving, negative or positive, is rendered in the Mandalas by the picture of Male Deity in contemplative physical union with Female Deity, a design probably taken from the worship of Shiva.

Shakti Laya Yoga, therefore, cannot be thought of as lust in any sense of the word in which our puritan-repressed natures think of sexual pleasure. Such Yogins who are real researchers would as lief suffer pain as experience orgasm. Both are violent discharges, psycho-physical catapults on which to launch the psyche daringly in flight out of the physique. Possibly Gnosticism attempted to introduce methods of this sort into the Levant. Dr. Inge, with many other European theologians, has spoken of Gnosticism's "oscillations between fanatical austerity and scandalous license."[2] These may have been no more than alternating attempts by pain and pleasure to enlarge consciousness. The attempts, we may assume, had to fail when introduced among peoples who had already undergone the Puritan Revolution, and to whom, therefore, all sexuality had become subject to the violent oscillation between craving and aversion. The real answer and criticism of this Yoga seems to lie in the fact that its practitioners are aiming at a new stage of evolution by attempting (through inverting the sexual act) to produce individual physical perpetuation, instead of perpetuation of the species, procreation.[3] That

2. *Christian Mysticism*, p. 82.
3. The Yogin can also maintain, though this hypothesis is even more alien to western thought than that of physical immortality, that his desire is not for physical immortality but only for so long a life as may permit him in one lifetime, and without the waste of time caused by rebirth, infancy, childhood and senescence, to attain the enlightenment which releases the soul from further reincarnations.

is probably the reason why Yogins of this sort were drastically persecuted. Whether or no this is a possible evolutionary development, that man should recapture the physical immortality of the first unicellular forms of life, we must suspect even were it achieved, it could not be the real purpose of life, for however remarkable the powers and new senses of this body, any localized vehicle and instrument must be, not an aid but a limitation to a fully enlarged consciousness, the consciousness which has attained enlightenment. As the philosopher MacTaggart remarked about the liberation of the discarnate: "Though while in the house I need windows through which to see out, when I leave the house I do not need to carry a window with me in order to see."

The question, however, still remains: Can intense physical pleasure be used to widen the aperture of consciousness? The answer lies probably in the answer to the complementary question: Can pain be so used? Mortification has not been found a satisfactory means. The real objection to both these methods is that neither of them penetrates to the source of evolutionary energy, each attempts to use a secondary manifestation and not the primary force of which they are, indeed, little more than symptoms.

If anything is disclosed by this attempt to set in chronological order these efforts at enlargement of consciousness, it is this: that man gains not merely increasing understanding of what it is that he desires and where that goal may lie, but also he realizes with what precision and delicacy he must release the unfolding flower from its sheath. There is needed an extraordinary subtlety of touch, a tactful vigilance, if the two converging aspects of consciousness, the "stalagmitic" growth up from the subconscious and the "stalactitic" exploration down by the analytic mind, are to make contact. If we are to advance through the evolution of consciousness beyond self-consciousness, to reunite consciousness now divided into the self-conscious and the subconscious, and bring about a new and single quality of awareness and being, there is required of us a peculiar and rare quality of attention. It must be intense but without effort and strain (see Chapter XI) . It must be that over-plus of evolutionary energy, which we call curiosity, and the German tongue calls *Neugier,* the new appetite, raised to what Plato called illuminating wonder. Indeed, because we are now so deeply prejudiced in favour of the assumption that it is only possible for one real experience to be presented to us, and that can only be the world of common sense, the necessary state of mind required for enlargement of consciousness must be one which is even without expectation. We are waiting for an experience of which we have no conception from the past, and that being so, any clear expectation must distort or even completely inhibit a radically new awareness. That is why so practical an explorer as Albertus Magnus, though practising in a semi-anthropomorphic religion, can rule so strongly that the experimenter must repeat to himself that what he needs above all things is above all that he knows, and that to rise up into the darkness of the mind is the one way to contact a further range of reality. This, as every reader of the high mystics knows, is their invariable experience and constant advice. So we can decode their negative language, realizing that they were reaching out toward experiences of consciousness beyond individuality and outside Time. As we have seen in the last chapter, the phraseology of the mystics is inadequate and often misleading. Yet the attitude of mind is essentially scientific for one of the most fruitful disciplines in scientific research is expressed in the aphoristic warning: Beware when you find what you are looking for.

Such alert passivity is peculiarly hard to maintain, but unless we can make a definite technique of this subtle process, our evolution will not continue. What we have to do, if we are to attain constant sight of that larger universe or Frame of Reference (that full and real world without knowledge of which even our normal rudimentary morality is sanctionless) is to know not merely how to dilate the aperture of consciousness but how to hold this enlarged awareness. While, then, we know that sporadic extra-sensory perception is possible, it will only and can only become something more than an anomaly if we can summon and maintain at command the profound insight of which it is merely a by-product.

CHAPTER XI

INDIVIDUAL TRAINING

THE discovery made by the later mystics, that expansion of consciousness cannot be forced, but must be negotiated, that it is in fact, of all explorations and discoveries, that requiring the most tact, delicacy and intuition, indicates the path to be followed to-day. Indeed, as at this stage the two different, ways of apprehending—the integral and the intuitive, and the analytical and experimental—converge, so we should expect that two techniques for reducing our apprehensions into comprehensible form (the technique of Art and that of Science) should also give rise to a third form, combining these. It is in the main a new type of attention. Attention may well have grown from a rudimentary tropism—an attraction as blind almost as that of the iron filing for the magnet and the force which drives the moth into the flame—up through an interest which could increasingly resist the urge to interfere. This detached curiosity, which no longer is content to end by affecting its object, then begins, as it were, to retract its focus until from being engrossed in the object which first excited its interest, it ends in focusing on itself; it becomes wholly curious not as to the nature of the object, but as to the nature of its interest in that object. This is what Mr. J. W. Dunne in his *Experiment with Time* has called observing the observer.

The first attempt to make a scientific investigation of the psychophysical conditions which attend upon and favour expansion of consciousness is the experiments of Professor J. B. Rhine at Duke University, North Carolina, in controlled extra-sensory perception. He has established the following important facts, all of which confirm the earlier "field naturalist" observation in this subject. In the first place, the faculty only outcrops, spontaneously and unmistakably, in stocks where its presence has been recognized in earlier generations. Perfectly clear cases of Telepathy and Clairvoyance are practically never found save in families where an hereditary tendency has been remarked. In the second place, however, experiment shows that the faculty is much more widely distributed than has been supposed. This discovery has led to a third one: It is highly probable that the clearness or obscurity of extra-sensory perception has little to do with apprehension and much to do with comprehension. In other words, it is not the strength or weakness of the impulse, but the thickness or thinness of the "threshold" between the two layers or aspects of the mind, between the self-conscious and the subconscious, that makes this method of transmission clear or faint, unmistakable or ambiguous. It may well be that everyone has such a power of apprehension but "this muddy vesture of decay," that strangulated condition of consciousness which gives us the conviction of being completely separate individuals, makes any conscious communication between the two levels impossible. The fourth discovery throws further light on this condition. It also explains another small but important point. It is that any attention by the self-conscious ego to this other process of awareness always greatly lessens and very frequently inhibits its functioning. It is this interference by self-consciousness which has made accurate, repeated laboratory experimentation with extra-sensory perception peculiarly difficult. So, either such upcasts of the deeper levels must take place when no surface attention is being paid or they must come through woefully mutilated and distorted.

These four discoveries may be summarized by saying that while all human beings possess (or rather are attached to) a depth of the mind, a submarine continent, of which the individual ego is a coral atoll, all individuals rise from and are planted on an ocean floor through which they are fundamentally united with each other, few can pass up and down the column, few can act as psychological amphibians. It needs a special training to be able to realize that immense underlying unity. Our rooted assumption of our individualism and our analytic method are aspects of the same phase and epoch of consciousness. Hence when we strive to analyse and by that means to control extra-sensory perception, we arouse our self-consciousness and that inhibits the manifestation of the very faculty which we would study and command. Dr. Rhine has as yet not been able to construct a positive method[1] which would aid the mind to function integrally or with expanded consciousness. He has, however, been able to establish satisfactorily the conditions under which this higher functioning is gravely handicapped.

As we have seen any self-consciousness, and so any outer interest which arouses self-consciousness, always interferes. What is of even more interest in this inquiry is the fact that any opiate—even of such a mild and sedative character as sodium amytal—inhibits extra-sensory perception. This would show that all drugs which have been used in religious history, from alcohol and fly agoric, probably only aid the subconscious by destroying the intense inhibition exercised by self-consciousness. As alcohol, though a depressant, may seem to act as a stimulant where it removes the inhibitions of shyness, as radium used in cancer treatment destroys the malignant cells more quickly than the healthy (but if left will destroy these also), so the drug may destroy false consciousness before it deranges the true, and distract a mistaken attention before throwing out a real awareness.

This discovery suggests another and larger one. Extra-sensory perception is, it seems, an advanced form of consciousness. Why we have considered it to be a backward and even primitive faculty (when we have allowed of its existence at all) is due to two reasons. Firstly, because it functioned more freely before the rise of acute self-consciousness, in the Modern Age. That acute self-consciousness, through isolating the individual, makes him assume that the world is alien and mechanical, and makes him also believe that only by increasing analysis can he understand anything. Secondly, since the rise and triumph of mechanistic, atomistic analysis, the integral function and the intuitive mind have hardly ever shown themselves except among peoples who have not been affected by individualistic consciousness or in such persons whose analytic self-consciousness is so weak that they cannot inhibit (and also cannot guide or control their apprehension of) the deeper and wider consciousness.

1. Indications of the right conditions for favourable functioning are, however, being carefully collected. See the Experimenter-Subject Relationship in Tests for Extra-Sensory Perception. Journal of Parapsychology, Vol. 2, No. 2, June, 1938. See also Upton Sinclair's "Mental Radio" and R. Warecollier's "Experiments in Telepathy."

Finally, Professor Rhine's important discovery that Telepathy and Clairvoyance are in reality only two aspects of a single faculty, extra-sensory perception, indicates that the mind when so functioning is not merely transmitting from one individual to another—practising, to borrow Upton Sinclair's phrase, Mental Radio—but is actually contracting a general field of consciousness, in fact, that extended consciousness into which it is humanity's aim to emerge.

We can say, then, that study of extra-sensory perception itself suggests not only that evolution is continuing in consciousness but also that after an intermission caused by the rise of acute self-consciousness, man is again about to enter a phase when his consciousness will enlarge itself further by a new technique of attention. Evolution is an alternating current, a tidal flow and now, after a regression, when consciousness at least added to itself an unprecedented acuity, it will again advance. It may be that as classical and romantic ages alternate in art and periods of classificatory activity alternate in science with periods of exploratory empirical departure, these are symptoms of a profound oscillation of consciousness itself, as it rocks to and fro in its spiral ascent.

What is clear is that in this new phase we have a development of attention, and that this is a double development. There is a change in what is attended to and also, what is certainly of not less importance, a change in the quality of the attention itself. We may say that man's primary form of attention was engrossed interest, his secondary form, detached and alien curiosity, and his third, which synthesizes the two earlier forms, once more an intense interest in which he has not to command or coerce his attention, but in which also there is now lacking any wish for animal satisfaction. He holds his attention and it is held by a realization that himself and That, object and subject, are polar, and through their interawareness, reality is created and consciousness completed.

If then, the whole purpose of our evolution and existence is the advancement into a higher and more extended consciousness, if we have now reached a stage when that evolution must, to achieve its purpose, for the preservation of civilization and for the development of ourselves, be intentionally pursued, it is clear that nothing else so much matters as the attaining to such a power. If we fail to advance —achieve this enlargement of consciousness—then we are ourselves frustrated. Our descent to Avernus may go by one path or another, slow or precipitous, but it will reach the same goal—the destruction of a species which balked at undertaking further evolution. If we are sensitive and self-centred (that introverted type which is produced by a sentimentalizing individualistic democracy), then we degenerate individually. The thwarted energy within us, which should have gone to our further evolution, will disintegrate us through increase of sensualism, by an ever more febrile eroticism; through increase of algesia, by hypersensitiveness to pain; through that overcharge of awareness which, not permitted to transcend the body, causes, in a degenerative succession, hypochondria, functional derangement, and finally organic collapse. We must recall the fact that diseases of the central nervous system are on the increase while what may be called specific diseases of the psyche are making ever more significant advances. The "hospital hours" spent in the United States on mental disease are now annually one hundred and seventy-three million, against the hundred and twenty-three million spent on all the other diseases put together. If we are active and extrovert, then we degenerate by mass neurosis. Frantic at the realization that we can discover no objective goal for life, we declare that our group, as it is or was in an imaginary past, is life's supreme manifestation. We make this delusion a persecuting faith, and with an epileptic energy attack all other social varieties, so as to make ours predominant.

This social degeneration is inevitable, even had not the mass neurosis of balked extroverts added fuel to the smouldering flame. For if we fail to achieve enlargement of consciousness, our societies are doomed as certainly as ourselves. Our culture and our civilization must themselves collapse. Our culture must wither for it can have no assurance of values. Our civilization denied sanction, must perish by that violence which it must employ, precisely because its values lack sanction. In brief, we must restore our general psychological circulation not merely to advance but if we are to avoid collapse. For the strangulation of our individualized consciousnesses has now become so complete that we are mortifying.

Yet such suggestions as follow will quite probably be rejected at the outset and without even a cursory examination of their actual feasibility, on the grounds that they are altogether too drastic. Such preparations, it will rightly be said, are only justified if civilization is going into liquidation. Things are not as bad as that. True, we are in difficulties, difficulties which rapidly grow graver; but when they have become sufficiently acute, the isolationist democracies and the privately civilized will all unite and become world-minded. If the League of Nations cannot be resuscitated we shall at least extemporize an *ad hoc* collective security against the aggressor of the moment before it is too late.

Such objectors to the idea that an order of avowed intentional living is the only adequate, immediate answer to the actual state of affairs, call themselves realists. Yet all their arguments unfortunately are obvious wishful thinking. Every month civilization is retreating farther from collective security as the international anarchy spreads. On what then, base the hope that mankind, as it is at present organized in sovereign states, governed by men specifically chosen to defend those states' interests against the rest, is about to reverse its secessionist movement? Five years ago it was possible, twelve years ago, probable. At those favourable times it was, nevertheless, not done and, it must be noted, the fact that five years ago there was a graver threat of collapse if the collective security was not obtained, did not make it an easier task, but less possible than it had been in the more genial and more hopeful political climate of Stresemann, Briand, Kellogg, and MacDonald. The only realist attitude is to see our situation as it is. Then we perceive that every effort to remedy it comes too late, would have been an adequate answer to conditions five years ago, but is now not a contemporary reply to the actual forces and the enforcible demands.

The truth is, the democratic sovereign states have lost the initiative. They are incapable of retaking it because they do not wish to see things as they are, but as they wish them to be, as they were when the democracies were the great powers. There actions are therefore all stultified. They only lose ground in spite of making sacrifices, because the sacrifices are no longer effective. We are, no doubt, now swimming with a growing sense of desperation toward the shore. In our anxiety we might redouble our speed. It will make no difference, for we shall only do so when and because the outgoing undertow has been trebled. We must repeat, this is not due to any sudden worsening of human nature, though it may seem so. Its fundamental cause is the momentum which our huge societies gather, once they start moving. An armament race cannot be stopped by the time that the man in the street sees war at the corner. The specious present, that moment when things happen so fast that it is physically impossible to prevent them, is for most individual men about a second. Within that time if a car cuts across his car, the driver is as helpless as his engine.

The specious present of the sovereign anarchic states is probably five to ten years. Within that "instant," whether they are democratic or dictatorial, they are as automatic as avalanches. It is that fact, as much as any other, that renders ineffective every attempt to make wise and foresighted the policies of the great powers. We can never swing these huge ships out of their courses within the time permitted us, the time between the moment when the shortsightedness of the ordinary citizen perceives the oncoming peril and the moment of actual collision. By the time he is aware of danger, the disaster is inevitable. Our first step is to regain mobility, to recover initiative. Only a new order of living can have the agileness to act with sufficient speed. The dalliers, the impractical idealists are those who still wish not to do anything drastic and so cry "be practical: get the governments to start all afresh." No more than could the dinosaurs, can the giant dynasts, whether they call themselves democracies or dictatorships, prevent colliding, and goring each other to death.

No cost, therefore, can be grudged if we may attain this new step in evolution. It is necessary to emphasize this because the price, as in all distinctive and decisive evolutionary steps is undoubtedly large. The newer the next stage, the more completely it requires an abandonment of the old. We cannot attain to it by any hasty movement of the surface will. A growing tree can be broken by a sudden gust. Its shape of growth can only be changed by slow and constant pressure. As the artist must wait with unstrained attention until the tide of inspiration rises, after which he must ride it to his goal, so to-day, those who would forward evolution can neither command nor neglect the inner drive. Coerced, it recoils; unattended, it dies down. The whole of life must, therefore, be directed to this end. Wide room must be found for watching and learning the flow of this current and how, when it sets, to take it at the flood and make the voyage to new levels.

The learning of a new technique of attention and making the achievement of that attention into the co-ordinating and mastering interest of our lives is then essential. Nothing less will serve. To it everything else must be subordinated. This task divides itself into two parts. There is first the actual mental process of higher attention, contemplation, that intense effortless activity which in absorbed interest experiences a consciousness transcending every personal feature, and makes contact with the unifying awareness of which every individual is only a strangulated extension. This clearly is necessary for a double purpose; not only because it raises the mind to contact with that higher degree of reality whence may be obtained the non-violent powers, but also because only a mind which has obtained and can retain such an experience can have transcended its ego. So, without the compulsion of threat, the urge of reward or the allurement of recognition, it is able to work co-operatively with others.

There is, secondly, the necessary matrix or setting of such contemplation, the whole daily life of ordinary consciousness out of which these higher states crystallize. Before, however, going on to discuss the essential conditions, economic and social, in which this enlargement of consciousness may be pursued, it is necessary to make clear what is the actual psychological or psycho-physical procedure whereby it may be attained. The very words whereby it may be described are indefinite. We speak of extra-sensory perception and many will say, what is this but mental radio, a free non-instrument telephone? True, the enlargement of consciousness has shown itself in this queer anomalous way, but as has been pointed out by Professor Broad, such was the history of electricity. It was first played with, and long known, as an amusing minute anomaly whereby powdered salt could be made to cling to pieces of amber. From such childish researches, however, grew the knowledge that the substance of matter itself is merely the balance of electric charges. The minute exception turned out to be the manifestation of a vast underlying law. So with extra-sensory perception, it is the small incongruent outcrop from an underlying basis of being, otherwise unapprehended. In extra-sensory perception we have that anomalous hint which, as Einstein has said, breaks the smooth circle of a smaller law and indicates that outside it lies a vaster principle. We must not then dismiss this thing because its laboratory indications are minute and inexplicable. These indications give us a handhold, however narrow, whereby we may approach scientifically the enlargement of consciousness and attain to what may be called the third Time Sense. The first is that of the animal, without reflection. The second is that of the ordinary human being. The third is when consciousness has rid itself of its involvement with past and future. This, however, cannot be attained simply by intellectual conviction. It is a slowly acquired training of the whole consciousness.

When, then, we have applied the findings given above from Dr. Rhine's work, we must add to them some further disciplines. Intellectually, advance will depend on "Right knowledge"; the first of the Buddhist eight categories which we shall find increasingly apposite. We must have discovered, as a matter of our personal experience, that the ego, the individualized self-consciousness, is not an end but a phase in the development of consciousness, and that time is that distortion and misapprehension under which individualized consciousness perceives reality, and so finds itself involved in inextricable conflict. Pyschologically, advance will depend on Right Action. That is, the bringing of the constant pressure of attention up to the limit of awareness. This is what is called in the West contemplation without images —in the East, samadhi—an intent alert passivity and vigilant openness, when the self-consciousness becomes aware of being exposed to and radiated by an intensity and range of consciousness completely beyond any physical experience of the self and, therefore, because of its very vividness, being beyond words.

Such an experience may be compared with the experience of the amphibian, when he first left the water, and to describe it to those who have not known it is as though the amphibian tried on returning, to tell fish about the world of air. Returning from it, many, such as Blake, retain for a while something of the enlarged intensity of its timelessness, and then the visible world, less distorted and blurred than is usual for us, appears with a keenness and a light which is of almost intolerable intensity, poignancy and joy.

Further, it is hard to go on giving in exact words a description of the state of mind which must be "held but not crushed." Fortunately, however, this general condition of mind, which must be maintained in meditation exercise for enlarging the entire consciousness, has been found empirically by those who had to teach insight into unfamiliar art. "Alert Passivity" was the phrase which the great art critic, Roger Fry, coined for that attitude and frame of mind which is necessary if a radically new vision is to be acquired. Nor has this discovery of the technique of newer and widened awareness been confined to the teaching of art students how to understand and take in contemporary art. Those who have had to teach physical techniques such as correct golf swing, correct voice production, correct physical carriage and right use of the eyes[2] have all found that they had to begin by teaching the mind to relax, to get round its false assumptions and wrong efforts, for in a materialistic world the mind cannot believe in itself and its powers and because that materialistic world is a "projection" of the strangulated ego consciousness all the efforts made by that ego to remedy its distress must only make the condition worse, whether it is eye strain, internal ulcer, voice production, or golf swing. These particular and proved techniques must and can now combine into a single embracing procedure and training. For, as we shall see, when dealing later (page 263) with the new technique of "Non-Violence" that just as this technique has been discovered and is already in successful use empirically in half a dozen therapies, although in its general social applicability has not yet been

---

2. See Dr. Jacobson's *Progressive Relaxation* and Dr. Bates's *Sight Without Glasses*.

realized, so it is also in this case of self-training. We have to see all these methods of specific reliefs and cures by "effortless effort," unstrained, unstrangulated awareness, as particular and partial illustrations of a therapy which in its entirety will remake and advance the whole mind-body.

Such states of rudimentary enlargement can be invited if not induced, by the constant readiness to welcome them, by periods put aside daily for such deliberate attention and enlargement, and thirdly, by employing during these periods such psycho-physical focusing as extended respiration and concentration between the eyebrows. In this way complete, unstrained, intense attention may be sustained and a clear awareness be obtained of a timeless state of consciousness beyond the ego's experience.

We must now turn to the second part of the task, the matrix or setting essential for such exercise in the growth of consciousness. As has been said, contemplation cannot be commanded but it can be made impossible by nearly every ordinary way of living. Appropriate diurnal living is the field in which sustained meditation alone is possible and out of this regular, sustained meditation, contemplation will suddenly precipitate itself. Further, though the moment of its manifestation cannot be commanded, it is possible that if adequate preparation and clearance are made, it will certainly appear with growing frequency—the earliness of its appearance generally bearing a fairly close relationship to the thoroughness of the preparatory arrangement. There must be reserved large space for its manifestation, for its intensity also bears a relationship to the time put at its disposal. If that time is scant and encroached upon by close-bordering engagements in irrelevant or contradictory activities then, like a bird which hesitates to settle in too restricted and overshadowed a space, the higher consciousness delays to alight, we grow fretful and anxious in waiting and this, in turn, further disturbs it. Even should it alight, the sense that it can only stay a short and stated time disturbs the state by making it (if we may use anthropomorphic language) realize that it does not come first, it is not the end of our being, its presence is not the state which we wish above all to be constant, enduring, exclusive, but that, instead of being the end, it is merely an interlude or a means to making more possible the ordinary living, which it should transcend in us.

The very shadow of any hastiness or restlessness is as fatal to this state as is sloth, for both show that the mind has not attained true interest. Hastiness is an impatience generated by the feeling that the present life is slipping away, and that if this method is not going to yield early results one must hurry back to the old life before it is too late. If this strain, which is fatal, is to be transcended (as it only can be transcended) by overmastering interest, then this study of enlarged consciousness, and this exercising of the psyche in experiencing it, must come first. It must dominate interest and have all the time it may require given to it. It is the purpose of living and therefore the one activity which must not be rationed. We must be glad when it takes of our time and interest increasingly, and when all other interests become subordinate. For this is evidence that, at last, the balance of attention is shifting from now to then, the centre of gravity of the mind is moving forward from the common sense life of irrelevant amusements and distractions to the life of complete relevance, of timeless being.

It becomes clear then, that much of the strain, conflict and exhaustion which attends the ordinary intentional effort to dilate consciousness and which frustrates that effort, is due to distraction of attention. The mind is under a strain because it is, as a matter of fact, attempting to attend to two things at once, nor has it decided which of these interests is primary and which ancillary. It has often not even asked whether these two levels of interest conflict or complement.

In short, it seems clear now that a higher level of consciousness will not emerge unless we plan all our living so as to permit that emergence. It is not difficult to understand why this should be so. The higher consciousness transcends the ego; the ego is a transitional stage, a husk. The present level of consciousness, of interest and attention, is almost wholly directed to preserve the ego. "We only live once" is a common motto and to make a life which may be worth while and have satisfied the self when it is over (almost a contradiction in terms) is the aim of "every sensible man." Dealing as we are, and cannot escape dealing, with a creature of body, mind and spirit, we must modify in every favourable way spirit, mind and body if the new type of consciousness, tentative and vague as it must be in the first stirrings, is to become fully formed.

It would seem, then, that the forwarding of the evolution of consciousness, the aiding of that further emergence of the psyche, requires, immediately, the constant practice of a certain method of mental focusing and this constant practice, in its turn requires, if it is to be successful and without thwarting strain, to be set in a complete way of life, a way which is centred on the exercise of this focusing, a way in which every activity is subordinate and ancillary to the focusing.

When such extensive requirements and radical modifications of customary living are said to be necessary, many may withdraw, using as an excuse that such a change of life for the sake of attaining a state of mind is merely escapist. The man who really cares for his fellows lives in the world, we shall be told. It is necessary, therefore, to remind ourselves that this contemplation which is to be made the ordering purpose of our lives, is not an individual, withdrawn, ingrowing, hypnotic pleasure. It is the deliberate attempt to continue evolution which is otherwise balked and arrested. It is a purposed plan to lead us out of our present impasse, to which unconscious growth and uncoordinated discovery have brought us, into the next stage of evolution, which is a higher degree of consciousness apprehending a larger and more relevant reality. What is clear, even to those who have not reached this next step and sighted the way out, is that we certainly cannot stay where we are and as we are. Our individual selves, our culture, our society and our race—all are profoundly unstable and rapidly becoming more unstable. However we try to escape it, we come back to the same point: we must go on or collapse.

It is, then, those who will not so recast their lives who are escapist, fleeing into social activity to escape thought, to escape the question, "Does my activity really solve the problem Life has set me or is my business precisely in order to escape from answering the problem which confronts me?" This extension of consciousness can alone solve our present problem. This problem is threefold. First it is psychical. As Havelock Ellis, who spoke with peculiar authority, has said, "There cannot now be natural sex—self-consciousness has made it impossible." "What is true of this intense and profound manifestation of human life is true of all other manifestations of our consciousness. We have seen that that other intense and profound sensation pain is also in a significantly unnatural condition. All our reactions are wrong: all our efforts only make our morbid condition more acute. We are all psychically unnatural for, though many manage to be their own warder and so be let at large, none of us dare let ourselves go. We all have to practise iron repression or be repressed. That is an unnatural and dangerous condition. Hence the steadily rising number of neuroses and psychoses. We are balking ourselves—and for what? For nothing worthy of a free man's worship. Democracy can only offer the pseudo-ideal of respectability and for this it demands that human beings deform their impulses as drastically as did the medieval morality, which at least aimed at making the subject a hero or a saint. We have already seen the beginning of the end. The futility of a Liberal regime based only on mutual self-interest, its only goal the physical satisfaction of individuals, who know that their individuality cannot be an adequate end, has led to Tyrannies, where the autocrat and his worshippers welcome any group action, however homicidal, because it makes the individual forget his inevitable futility.

We require, then, first of all a channelled release for this dangerously thwarted energy. We require to know that there is a further life ahead—not another world but a further growth of consciousness and an evolution of awareness into which by intentional effort we might merge. The thwarted energy can only be satisfied by such an extension, the puzzled consciousness can only be answered by such an explanation. We shall only check the spread of neurosis, first individual and then massive, first psychological, then economic and political, by discovering this, the only true and adequate outlet for man's intense creative force.

That being so we must find such a method of releasing the mind. So only may we adequately "cleanse the stuffed bosom of that perilous stuff which weighs upon the mind." We have to find, use, and train all who so desire, in the practice of a method of focusing the mind without strain, on its highest levels. To-day, in our way of life, we are like men who might be driven by some sadistic tyrant to focus only and always on objects within two inches of their eyes. The intolerable strain can only be relieved by permitting the eye to focus on distances adequate to its range. The world to-day, being one of grotesquely limited range, we must regularly withdraw and relieve the unbearable restriction by gazing at a satisfying perspective. So and so only will be able to act creatively. No day must be without this essential therapeutic extension. Otherwise the compression of the short-range world will permanently and fatally cripple and distort the spirit. As Irving Babbitt truly said, "Everything will be found to hinge finally on the idea of meditation. This idea has suffered a steady decline in the Occident, along with the transcendent view of life in general… Yet it is not certain that religion itself can survive unless men retain some sense of the wisdom which may be won by sitting in quiet recollection."[3]

The exercise of shifting the focus of attention toward extending consciousness is then immediately necessary if we are to gain sanity in our individual lives—to check the spread of neurosis in the remaining free countries and to cure the mass neurosis in the countries, which, in order to escape from the futile life of the self-conscious animal, are worshipping a fictitious national consciousness. It must come first. When a man is gravely sick—especially if it is of a highly infectious epidemic—the first need, taking precedence over everything else, is recovery. No time must be grudged as no money would be spared. Indeed, it is when recovery has set in that time may be found for other activities, for nursing others and running hospitals. When a cure of an acute condition is being begun, any time demanded for that cure's establishment must be provided, as much for others as for the self.

We cannot put right either our economic or our political systems until we cure our strangulated individualities. The first sign of returning sanity, the first symptom that we are attaining the essential objectivity toward our acute problem, is when we begin to realize

3. Irving Babbitt, "Buddha and the Occident."

that our societies are projections of ourselves and that our first radical contribution to a new social order and a conscious civilization will be made, and will only be made, when we undertake the reintegration of ourselves and set ourselves to continue our evolution by enlarging consciousness. Once that is attained, once we have a real psychology, inevitably there will appear a sane economics and a worthy politics.

Therefore, we must for every reason, for individual integrity and sanity, for economic continuance, and for political order, start with the deliverance and ordering of ourselves. Nor need we feel that this is indefinitely to postpone attending to the acute economic and political problems. As we shall shortly see, the moment we decide really to attend to a real psychology, we shall find ourselves driven to creating a sound economy and to demonstrating the only adequate policy.

For every reason, individual and social, we must, therefore, find time for beginning at once, and on an adequate scale, our development of consciousness. The trainee must set aside an ample margin of time, for at the beginning his efforts will be so clumsy that, feeling he is making no progress, he will start to worry about the time at his disposal. This anxiety will further inhibit his practice and, involved in a vicious circle, any advance will come to a standstill. Two clear hours, one in the morning and one in the evening, is probably the narrowest allowance with which a beginner should provide himself. This time must be reserved—whether benefit is felt to come of it or not. The deeper mind will at last recognize that these periods are being held for its emergence and it will emerge. As Lao Tzu remarked, "muddy water let stand becomes clear." Without silent recollection the mind cannot recognize its own unity, still less the wholeness of that into which it would expand.

Though, however, this period is a minimum, and though less time will probably show no result, being found to be below that critical threshold under which no true recollection and expansion can take place, nevertheless, it is immensely difficult for most people to reserve such a large period. Many begin energetically. Their surface will and the novelty of the experience keep them going for a few weeks. By then they have probably had a slight experience of preliminary shift of consciousness: they have actually realized what it might mean to experience frequently, and even to maintain, a higher awareness. Then (for this advance is certainly alternating and tidal), they find this hint of awareness dies out. The will tired, the novelty gone, they easily persuade themselves that they are only practising a futile, time-wasting, self-illusion. They give up. Sometimes, in moments of discouragement with the old irrelevant life into which he has relapsed, the man who has given up will plunge back. Impatiently snatching up the problem again, he will attempt to face the issue. This petulant effort only strains the delicate deranged valve-mechanism of the threshold and jams it more firmly. The man experiences no relief, only growing distress and exasperation. The whole subject becomes for him a painful and futile pursuit.

If, therefore, enough energy is to be found to reserve time for these exacting exercises, and so prevent the serious mistake of their being abandoned, it can only be—as has been said—if life is rearranged so as to make this possible. It may be asked here: "Is this then simply a luxury for the rich and unoccupied?" This question will be answered more fully later. At present, at this point, when we are considering only the individual working by himself, it may be said that the greater the cost of making such a change in the disposal of our time, the greater is the benefit. The deeper mind is well able to gauge and to reward the price that we are willing to pay for its company.

CHAPTER XII

GROUP TRAINING AND EXPERIMENT

(A)—Psycho-physical

THE first, easiest and most obvious assistance toward the individual's private efforts, and to saving these from intermittence or collapse, is by simple association with others undertaking the same attempt. We know that a small group of ten or a dozen tends, if it is like-minded and will meet regularly for meditation, to produce a favourable field. Individuals who are each attempting to enlarge consciousness should find such a number of similar associates. Then each of these, assuring the rest that they by themselves are attempting daily meditation will, in a weekly meeting of not less than an hour's silence, find considerable assistance in keeping going their individual practice. This is not much to ask, even of the most driven. The amount of time we waste is considerable, even among the busy. If for one day we keep tale of our loiterings and mind wanderings we shall probably discover that they amount to hours. Quite apart from our deliberate amusements, most of us fritter away a considerable part of every day so that we should scarcely be able to say what we did with quite a large percentage of it. Our parents were busy but easily found time for their church-going. This can take that place.

Simple silence is not enough. The mind can wander more completely when the body is still than even when it is moving. The Society of Friends has long used silence, and in the beginning with immense effectiveness, but in its attempts to avoid the dangers (or as a Buddhist would say, the fetters) of a theology, it failed to achieve a psychology. In rightly avoiding vain talks about the ends which are yet to be known, it failed to work out with precision the means through which the goal must be approached. Hence a tendency to relapse into anthropomorphism in thought, to decline in action into a conventional acceptance of the world of common sense, and to degenerate in meditation into reverie. It is necessary for those who would practise meditation to be quite clear about two things: the end and the means; goal and method. The Quaker says that he meets for worship—worship of God. Such a term very easily becomes anthropomorphic—a lesser individual praising a greater. The Quaker achieves that worship in silence. But what is the mind doing about that silence? Silence may be mental talk, mental wandering or mental concentration. As, then, there are two chances out of three that silence will not be superior to speech, a very clear knowledge of what is intended, and how it may be done, must be possessed by those who would use, with any results, this advanced and difficult method. Some definition is therefore necessary and that definition must be such as to prevent too narrow an assumption.

We must then define our end and goal as the expansion of consciousness. Such a definition leaves the way of advance open while it does not commit us to what, in our present knowledge, must be a premature and limiting definition of the goal's nature. We may define our means and method as such mental exercises in detached focusing of the mind as may lead to such expansion.

With these definitions we can now see how simple silence can be filled with productive psychological activity, activity above, not below, the level of the mental process which speech permits and assists. Nearly every description of the Yogic methods of mind training concludes by saying: These oriental methods are not suited for occidentals. Are we then to remain between two schools—between the Indian definite methods which work, but not for us, and the Quaker indefinite procedure which, because of its lack of psychological precision, too often relapses into anthropomorphism and increasingly fails to attract Seekers? It is clear we must combine the two. The time has come for another of these great syncretisms of two highly-developed but long-divided lines of thought, those syncretisms which give rise to a new departure of the mind. Yoga and Quakerism must be blended in an original synthesis. The specific contribution of Quakerism is a highly-developed community sense. We must leave that alone for the moment until we come to discuss the inevitable part which economics and politics must play in this our age's unique attempt to advance evolution by the expansion of consciousness. The specific contribution of Yoga is knowledge of practical psycho-physiology. We must, therefore, in this place consider how this knowledge may be modified and applied to our immediate conditions.

The first thing that Yoga teaches is a truth which we can certainly understand. It is that to change the focus of consciousness is difficult and skilful work. It does not happen by accident nor by simply leaving the mind open. A simple devotional attitude even is not enough. Every assistance must be used: actual surroundings, at the time of the psychological work, diet and economic circumstances. First, then, as to the conditions under which the work is actually done. Here Quakerism, in its attempt to avoid dead formalism, runs the risk of an equally dangerous casualness. "All life is sacramental" is a favourite aphorism with the Friends. This, however, is and can only be true of those rare spirits which have advanced to complete enlightenment by complete release from the self already in this life (Samdithakam Nibbanam of the Buddhists; Jivan Mukti of the Vedantists). For the rest, even of many who have been considered saints, there must be a fluctuation or at least a systole and diastole. The Quakers themselves acknowledge this in practice or there would be no need of their silent meeting. No aid can be neglected in helping the mind to rise above itself, so long as it is a real aid.

No doubt, we need in this matter to experiment. Because we know our goal and our means we may seek as widely as possible for anything which will aid those means. Unfortunately it is here that religion has been most timid. Unless early authority can be shown for any practise, it is shunned. "The idea of experiment and the idea of the supernatural have come to seem to us mutually exclusive."[1] We must recombine these ideas and we can, because we see they were driven apart owing to anthropomorphism. Anthropomorphism must, of its nature, always treat any idea of religious experiment as grotesque and blasphemous.

We need, therefore, to use any aid we may to "bring our mind in frame" for exercise in expansion—as we must now call it, and not

---

1. Irving Babbitt, *Buddha and the Occident, p. 78.*

concentration. There are several stages of release, and they must each be successively and completely negotiated, before entering on the next. By unskilful haste, by impatiently concluding that we have completed the preliminary detachments, we plunge at the deeper levels, only to find ourselves pulled back by earlier ones from which we should first have freed ourselves but which are still holding too much of our attention, as a diver, plunging over-board, might find himself returned to the surface by his air-filled clothes. The body and the mind must be successively relaxed before we can expect to attain natural and unstrained expansion of the spirit. Such a work as Dr. Jacobson's *Progressive Relaxation,* referred to on page 199, Chapter XI, shows three things which are very relevant at this stage: how slowly the body relaxes (time is essential to unlock these profound psychophysical tensions); how, as long as it is hyper-tonic, no effort of the surface will can free the entire attention; how, once this hyper-tonicity is relaxed, the health of the body increases, diseases cure themselves, sleep is at command and attention whole and free. Even the eye, one of the quickest organs to adapt to change, has in certain circumstances to be waited on. It generally takes some twenty minutes for night seeing to be attained and those too impatient to wait may take a wholly unnecessary fall; while physicians studying X-ray photographs and using the Fluoroscope, find it increasingly helpful to mask their eyes for twenty minutes before looking at the screens, lest their unadapted sight miss in the picture faint but highly significant traces, shades and discolorations.

In an actual weekly meeting of a dozen like-minded meditators, as time must be permitted to fix the outer limits, so time must indicate the order. Music is for us a great detensioning power. It should probably be used as a preliminary at all meditation meetings. Its pure pattern is superior to any words, however lofty, for words carry images, of which the expanding mind must free itself, and involve us in narrative, along which the mind strays. The Sufis, who are denied our music and confined to the Koran, still attain the detensioning of rhythm first by eloquent declamation of rhetorical passages and then by constant repetition in rhythmic strophes of the name Allah.

So much can be said about actual group practice of expansion.

The body and mind being brought to recollected repose, the spirit can press with skill and with increasing progress against its limits.

It is clear, however, that body and mind will not abate their disturbances unless, beside the actual hours of meditation, the whole current of life is in keeping with, and contributory to, that crest and growing edge of attention. That brings us to diet as a necessary contributory factor and setting of the actual psychological work. Diet is a wide word, and we must not let it be narrowed down to mere menu. Food has considerable importance and the experimenter in expansion of consciousness takes care that his nourishment aids his mind. The tantric Yogins teach that food must be light but the stomach must never be wholly empty, or it will distract the mind by movement, however slight, and no energy can be spared for this distraction. The higher the attention being aimed at, the more easily the mind in the preliminary stage is distracted, for as in sleep—which is the descent into unconsciousness—so in contemplation, which is the ascent into higher consciousness, we have to pass through a belt or phase of hyperæsthesia. They advise, therefore, a light meal every four hours. Diet also includes the whole planning of the day—sleep, work, meals, meditation. Every activity must be rationed so as to be menial to the development of consciousness. Nothing that could be a means may be permitted to become an end. Every tendency to addiction must be skilfully forestalled, and by drawing off interest, the uncoordinated subgrowth be atrophied. Addictions are more often symptoms of boredom—lack of vital interest—than of passion—an excess of animal energy.

Here we must ask, what place, then, is given to sex? It seems probable that all the vital energy which manifests itself in our vagrant sensuality is required for the high activity of expanding consciousness, and when meditation is of intense, effortless interest, complete sublimation is effected because the primal energy is wholly used up in this supreme limit of experience. Certainly craving cannot be felt when attention is wholly given, and in ecstasy an experience of such completion is known that after it the sensory release of the orgasm, as that of drunkenness, is recognized as being a hopeless parody. Though, therefore, Vaijroli may seem to offer a secondary path to those who say they cannot sublimate, if Right Contemplation, Samadhi (the words are the same in Pali) non-personal consciousness (ecstasis) is not only possible but the actual getting into the next evolutionary stage of consciousness, then surely we must aim at nothing else, and the problem of sex by this and by this only, finds at last its solution.[2]

Yet our world, lacking knowledge of ecstasy, is tending to thrust sex and alcohol and all "things that give a lift" on the starved consciousness's attention. It must be so. Here is this great unused energy gnawing at us to gain release. The materialistic, democratic civilization offers us only the ideal of respectability and the anodyne of amusement. No wonder that, with all further perspectives of exploration closed, the word "adventure," once connoting risk and exploration, sinks, in all such arrested societies, to mean nothing more than the hunt for sexual novelty.[3]

If, then, we would attain ecstasy and the freedom from all addiction which it entails, we need not merely psychological exercises and helpful diet. These will not be able to work freely, so encroached upon are we by the common distracted contradicting life of amusement, the appeals and advocacies of which have become increasingly more strident and urgent as individualistic society realizes more intensely its frustrated state. Beside deliberate method and diet, we need intentional economic circumstances. Our whole society is increasingly organized to keep its economy going by creating craving and constantly inflaming it. Our economics are based on increasing consumption, and when those who can pay for it can consume no more necessaries, we must make them buy continually more luxuries, otherwise our crazy machine "stalls" and we slump and crash. The only alternative our world knows to this is an armament race. It appears inexplicable that we should ruin and destroy ourselves when plenty is at last within our grasp. The reason becomes sadly plain when we shift our diagnosis from the economic symptom to the

2. Sex here is, of course, being considered not in its reproductive aspect but as it is now mainly considered, as amusement, distraction, de-tensionment
3. It is significant that the word boredom is not in Dr. Johnson's Dictionary, and that the mood is first recognized by men of the world in the pensioned and semi-imprisoned nobility of Louis XIV's Court, though it probably existed for monks with no vocation throughout the Middle Ages under the mortal sin, acedia.

psychological cause. Then we see, as clearly as Damocles, when he suddenly lifted up his eyes from his plate to the ceiling above his head, that this is exactly the economy which we should get from our present psyche. The ego, driven by greed and fear, will produce alternately an economy based on perpetually increasing the demand for luxury goods and one based on perpetually piling up arms. Either drive will end in collapse and it really makes little difference whether we go over the gunwale into the water on the port or the starboard side, by internal revolution or external war, by a hopeless conflict between the verticals (international war) or between the horizontals (class war).

(B) —Economic

This brings us to the second main consideration. We have seen briefly the outline of psychological elements required for expanding consciousness. First, the constant daily exercise (which by its demand on time already makes the ordinary conventional life difficult), and then the weekly meditation meeting with similarly engaged associates. This, of course, is to have reached the level of an essential Church. This intentional social organization can do something for the individual who becomes an organic part of it. It can begin to rid him of that constant anxiety about his security and his purpose. It can make him part of an unlimited liability company. It can spread his risk as well as give him confidence. But it cannot do enough. Such a "church" is only a small group of like-minded people and even though such groups might affiliate with each other, even if they should consist of people who had been wealthy (an improbability) they could not give their members the right freedom from anxiety, that freedom to attend wholly to the advancement of consciousness. For, however powerful a church, it is involved in the world of greed and fear. Indeed, the more economically powerful it is, the more it is involved in the present system which generates wealth through the forces of greed and fear, through exacerbating individualism. An endowed church, therefore, gains its security and gives its members security out of a tainted source. It attempts to leave men free from economic anxiety, so that they may pursue untrammelled spiritual ends, by buying their freedom with power which comes from economic compulsion. We cannot, if individualism is the real captivity and unity the only freedom, make men free by endowing them with the energy produced by slaves.

Yet the only other alternative has seemed to be the Indian ascetic's begging bowl. We cannot deny that as long as life is psychophysical, so long the psychological explorers, though they do not live by bread alone, cannot live without it. They cannot command that stones become bread. But the begging bowl is not really a true alternative to endowment. It may be economically more sound to have your plate filled every day with actual food than to be given a large number of dubious debts and told you may take the profits if you can get them (which is all that endowment with stocks now means). The beggar, however, depends just as much on the crazy greed-and-fear-driven productive system as does the wealthiest bishop or college president. He does not get down to any underlying rock. This means that his economic security is, therefore, no greater than the worldling's. What is worse is that for the sake of a precarious economic security these beneficiaries imperil what is to them much more important, their spiritual integrity. The begging monk or the college president are given their keep not for nothing. They are provided for because they give in return luck, merit, salvation to the individual who pays, or prestige to the corporation which endows them. The begging monk becomes part of the economic system which it is his life to transcend and to show all others how to transcend. Living on charity is not merely economically unsound: it is psychologically far more unsound. For the surplus from which the mendicant or the endowee is fed, comes from an acutely diseased society, and is given to these beneficiaries expressly that they may sanction and so secure the continuation of such a society.[4]

The one question which confronts such pioneers as may form an essential church of a dozen, is how their consciousness may be free to evolve: and so they may give rise to a new humanity and prove to others how this may be done. Such consciousness cannot be freed from distraction and even worry simply by economic security, regardless whence it comes. Even if the present economic system would and could give such pioneers ample lasting and certified provision (which it certainly will not and could not), such economic security, because of the world from which it is derived and the way it is extracted, would disturb the calm it was meant to preserve. A house built on a plague site, imperils its inhabitants.

A group of explorers, therefore, needs a third element if these explorers are to be able to advance. The first requirement was a way of living followed by each member, so that he may integrate his nature, reblend his fissured psyche and direct his wholesome nature in the direction evolution indicates. This is the individual psychiatry. It needs to be psycho-physical. The whole body-mind must be directed in its diurnal activity toward the advancement of consciousness. This leads to the second requirement—the association with others in this intense, constantly intentional activity. This is necessary because at the actual work and in the secondary and supplementary activity of all the rest of living, the individual needs to work in reciprocation. He needs, beside joint meditation, discussion on procedure and results and also the constant planning of conditions so as to lead to greater freedom and energy for the work, and greater efficiency at it.

When this has been stated it is clear that there is a necessity for an economic aspect of such an endeavour. It is equally clear that the economic aspect must not only be wholly ancillary and subordinate to the psychological, but must be completely independent of the present economic system.

Beside the precariousness of the present social system, a precariousness which makes it economically unreliable, it is also tainted with neurosis. Its economic instability might be suffered if it was not

4. The begging monk, especially in the East, justified his procedure by saying that the low food-producing type of man, by satisfying a monk's physical need, acquires the only merit of which he is now capable. Therefore the monk not only may—he is bound—to give the creature this chance. The monk is giving far more by accepting being given to, than he gets. This is to assume that Transmigration is taken as self-evident by donor as well as by donee. When this is not so, the argument is greatly weakened. For then the donor will only give for luck, in this world (hardly for pity to a man who is in fine physical condition and so could produce food for himself), and so be the worse for giving; or he may refuse to give at all, on the grounds that the monk is a fraud and cannot yield a magical *quid pro quo*; when the monk must work or starve.

a symptom of a deeper dislocation, the competitive and antagonistic manner in which the social system's constituents are functioning. Nor is this a condition which any individual and personal adjustments can remedy. The plain fact is that all actual sovereign states are sprung from and have been developed to maintain international anarchy; they are all secessionists from civilization and morality, for they claim to be above Law. As they have grown they interfere increasingly with the private lives of their constituents, compelling these, their members, to co-operate ever more actively in their criminal designs. The explorer who would advance consciousness has to consider, therefore, that as long as he accepts this society's charity he is taking part in its activities, activities which must frustrate the explorer's whole purpose. His purpose is to attain a consciousness which has transcended greed and fear and the whole strangulation of the ego, to emerge on to a larger, comprehensive consciousness and to show all others that they are part of this and may so live if they will, by emerging into the whole of being.

Nor is such a demand quixotic, a leap in the dark, an abandonment of proved securities, and the sacrifice, for a wild adventure, of circumstantial plenty and means. Our present economic system is so intractably out of gear that some radical adjustment must be made if the whole machine is not to "seize" and the system fail to provide any longer even material means—let alone be an object in whose service its members might find a worthy aim and end. We are beginning to understand that the same apparent paradox which haunts hedonism—that happiness can never be attained by direct approach but only as a by-product of some farther-ranged activity—also rules economics. Only by aiming at an order which transcends the economic and the material can we achieve fine physical conditions; only when we care for something quite beyond healthy living conditions, can these be secured; only when we care for and are devoted to something transcending humanity shall we once more at last be able to be humane.

As we are social creatures we must find company which is going the same way. This is not merely for mutual assurance and discussion of methods of meditation, but also for the strengthening of the methods themselves, through a joint exercise. Yet such a society or church cannot be without an economic aspect, and as that aspect must be one which, not only supports the body by making it free of "wage slavery," but also respects and puts in the first place the spirit, that economic aspect must itself be distinctive. It cannot be a mere parasite of the present system because the present system is rooted in that very psychology which these pioneers have to transcend. Hence, if evolution is to continue (for it cannot continue now unless we consciously co-operate with this, its next step, the evolution of consciousness), men who are forwarding that evolution must make for themselves not merely personal and private ways of life but also a new social pattern of living which permits and expresses their new type of consciousness. They must create a manifest social pattern of avowed intentional living in which the higher type of consciousness can function. They must form a community which has the three functions which makes a society an organism capable of creative growth—an effective psychiatry for uniting and expanding the individual, an appropriate economy for sustaining and forwarding the like-minded group, and an original policy—a message and a demonstration to the world that there is a way out of its tangle and a way ahead for life, if people choose to take it and pay the price.

It is now obvious that men cannot live much longer the life of irrelevant distraction. We are becoming too acutely conscious and aware of time. Any stimulant, however pleasant at the beginning, repeated becomes hateful. The life of amusement and leisure for all, will never arrive, we may prophesy, because of the demonstrably tragic evidence of its effects on the large minority who already have been exposed to it. We must know we are getting somewhere or we will go back to violence and distress. The intense and savage regimentation of the Totalitarian States is simply the contractive reflex of peoples too long exposed to private and general futility. The goose step is no more than an attempt to shake off that goose-flesh feeling.

Democracies and Dictatorships therefore both compel those who see that neither the present individual nor the present state can be the end, to show clearly not only the goal but the steps leading to it:

to live a complete way of life which has not only its right psychology but an economy which agrees with and is the outcome of that psychology.

It well may be a pity that this expansion of consciousness will not come about of itself, like the growing of one's finger nail. It might, indeed, save us trouble if (failing that) we could carry out this development personally and privately. Yet, if it is possible and is also the only possible evolution worthy of a self-conscious being, then no price is too much for a life which not only saves us from futility (for *that,* in dictatorial countries, men have paid everything: their liberty, their integrity and their lives) but also makes us a part of a really higher consciousness, more intense than our self-consciousness and embracing all being in its unity.

If, then, this life has to have an economic aspect, a social base, we may give ourselves to the planning of it with the less hesitation when we realize that if we do not make a clear and intentional way of social life, there will be forced on us a way of living which will be regimented and dictatorial.

Before we leave this economic aspect there is a specific economic question which must and can be answered here. To most people, unaware of modern technical progress, the thought of a community which has predominantly psychological aims, raises in the mind the picture of a society painfully inadequate to supply sufficient resources to permit sufficient time for psychological advance. The independence from our tainted, state sources of wealth is only won by exchanging it for dependence on a Nature which keeps its "adscripted" serfs at a physical slavery for starvation commons.

This relapse into toil-dulled peasanthood is, however, no longer necessary. Man need not choose between being a social parasite and an agricultural drudge. If it is necessary for psychological advance, if it is required for further evolution that we make a directly productive society, a self-subsisting community, then it is not economics which any longer questions the feasibility of such a plan. Such authorities as Borsodi[5] have shown that, with the rise of small power plants, a

5. School of Living, Suffern, New York.

specific community can with their help in sum actually live more plentifully—fewer work hours will yield them a greater sum of goods—than under the present economic system. Even if more means were the only end of a new small community, such a community would attain it better by means of small power plants directly operated by themselves than through our actual muddled mixture of mass production and chaotic, congested and expensive distribution. Quite apart from the gluts and slumps, the over-production and the failure of money-controlled purchasing power, which make the present system an anxiety nightmare, even if the system worked, it would not be superior to a decentralized method, which rationally planned groups and communities would work for themselves.

What then prevents intelligent people availing themselves of this way of deliverance? Nothing but the fact that society is, and must always be, based on psychology and have as its consequence and symptom of that psychology, an appropriate economics. We have tried to maintain that this is not so, that in fact the reverse is true, that economics is basic and psychology the resultant. In actual fact what we have is nothing of the sort. We simply endure a diseased economics inevitably springing from an evil psychology. Our society is founded on an individualism which is made to produce means by appealing to its greed and fear. No wonder its economics are crazy. The wonder is that they have worked so long. They cannot, however, be cured by attending to them only or first. They are symptoms caused by an underlying psychology which so expresses itself. Cure the psychology and the evil symptom will disappear, and healthy functioning, a sane economics will take the symptom's place.

This cannot too often be repeated, because our minds have become conditioned and fixated into seeing our problem the wrong way round and so, in all our efforts, Liberal, Communist and Totalitarian, we only inflame our disease.

If then we would have sane economics, we must have a sane psychology. In short, only an avowed intentional order can avail itself of this offer of a sound economics. Any other community based on unresolved egotism would fail to make function these new means of rational and ample supply. The uncured anxiety and greed would ruin the essential co-operation, as the uncured fear, suspicion and lust for power ruin Communism in Russia. Only those who have discovered the path of evolution, what is its next step and how we are to co-operate with that development, can know themselves as part of a self-transcending purpose. Only such as have this experience will possess the mutual social sanction which will hold a community together in an organic relationship. Such people and such only will have and cannot fail to have a sane economics, the sane economics which will give them the physical independence required in order that they may be free to advance to further consciousness.

The level of economic requirement, the standard of life, is then settled by the stage of psychological advance. The two must emerge together and balance. Even if these studies in small power plants for decentralized self-sufficing community-living had not been made, it is tolerably certain that the psycho-social insight possessed by men who have realized evolution's end and how they co-operate with it, would have given rise to the necessary economic planning and inventions. The converse is also true: the height of economic efficiency is exactly limited by the extent of psychological integration. We may make innumerable economic and physical inventions; they will remain useless or even turn malignant unless we have the psychological insight to render them beneficial.

We have now seen the necessary dispositions which must be made by those who would co-operate with evolution through the enlarging of consciousness. So difficult is this work that it requires a wholehearted interest and utmost care. It is a life work to which all other activities must be contributory. It is a new way of being conscious and therefore it must be precarious, tentative, often baffling and always absorbing. Hence, every distraction must be removed. Nothing must be done which does not contribute to this end. Hence, first the mental life, its reasoning and reverie, must be made to centre on this development. Sir Isaac Newton, once asked how it was that he arrived at his Law of Gravitation, replied that his mind never ceased thinking about the question. Then the physical life, its functions and habits must be made to sustain constantly the mind in that direction. More than we realize, until we watch, is thought initiated by some bodily prompting. This in turn, requires that the social life be arranged so as to be the expression and confirmation of the individual's mental life. The individual must, if he is to succeed without resorting to frustrating strain, be with others so developing, instead of with those who are lost or arrested, bewildered or automatized. The fatal social suggestion of the society which is without a purpose and even without a knowledge of life's meaning, must be exchanged for that of a society which both knows its goal and is purposely directing its every activity so as to reach that goal. Democracy throws the reins on the horse's neck, leaving it to find its way home. The swallow knows its world route back to its nesting place, the eel its ocean trek to its deep breeding bed, the ox knows its stall and the ass its master's crib, but individualized man does not, and as long as the ego remains untranscended, cannot, conceive his goal. Dictatorship, once in the saddle, puts a cross-bit in the creature's mouth, spur and whip to its flanks, and drives it to destruction.

There is a final reason why this work must have an economic and social aspect. The attempt to co-operate with evolution through enlarging consciousness must be of its nature the most public and general activity of which man is capable. There is no private salvation, everyone can attain to the state of general being if he will obey the urge to unity which stirs in him beneath his ego. The pioneers may have to go ahead, but they are not seeking a private escape, any more than the sailor who attempts to swim ashore with a rope from a wrecked vessel is deserting. It is essential that such pioneers should show quite clearly the steps which they are taking so that any who care to follow may do so. It is necessary, in order to mark those steps, that there should be an evident design of purposed living, a psychologically-based economy, as an alternative to the chaotic, individualistic competitive system of to-day, an alternative to that economically-deformed psychology, to that society resting on the assumption that man is only an economic man, an animal with nothing but physical needs.

We must make this life-size experiment in order to answer the supreme question which can be solved in no other way. At present we do not know whether psycho-physical evolution is over, whether further evolution can only be purely psychological. We have not given psycho-physical evolution a fair chance. We must not desert this world until we have proved whether, under the most intelligent methods and applying all contemporary knowledge, psychological, economic and social, a viable and progressive society could develop. It is possible that there might be a way of living whereby man's bodily needs and his psychical advance could both be provided for by the creation of a society in which the prime generating force was neither fear (as in dictatorial states) nor greed (as in the democratic states) but the realization of a unifying consciousness. Until we know for certain that men have rejected such a way of living, when it was presented to them as an actuality, we are not free of the hope or from the responsibility of demonstrating such a way to them.

It will have been noted that the system here recommended for helping the further advancement of evolution by the continued enlargement of consciousness has up to a point closely followed in modern terms the Eightfold Path taught by Gautama. "Right Understanding" is realizing what is life's purpose and aim, and what is our immediate place and work in that purpose. It aims at ever greater awareness and we must aid it by enlarging consciousness from individuality to super-consciousness. "Right Will" is the acceptance of this purpose and of our part in it. "Right Thought, Speech and Action" are making our mind-bodies in every momentary activity to be mindful of and to express the intention of our lives. "Right Livelihood" has needed, however, as we have seen, considerable explication and development. Gautama was peculiarly wise and careful in indicating that the way of livelihood which men are compelled to follow in a society functioning through greed and fear, must completely stultify their good intentions. "He who cleans a room in God's name can serve God fully." True enough if his work is an integrated part, however lowly, of a whole social activity which culminates in the vision of a complete evolution through which, as they strive, all will come to enlightenment. On the other hand it certainly is not true if the cleaner is making the room "empty swept and garnished" for seven devils the more amply to occupy. Doing one's duty and not asking questions, especially in the intensified integration of a modem, mechanized State, taking its sovereign part in the international anarchy, can be fatal. Indeed, our present pass would not be so deadly were it not that this simple doing of one's duty, the cult of efficiency, is precisely the factor required to transform the destructive delusions of the despot Dictators from the private dreams of paranoiacs to world-wide cataclysms.

Every student of history must be startled by the fact that in the Roman Empire, in Chinese Imperial history, and among the Indian dynasties, it is so often recorded that even when an autocrat was known to have become insane, still thousands of conscientious, high-minded officials were found ready to carry out his criminal orders. As a Russian historian has said, concluding his study of Ivan the Terrible: "I throw down my pen, not that such a monster as Ivan the Terrible should have existed, but that there were to be found hundreds of my fellow Russians to perform any of his atrocious demands." We have assuaged our alarm at these symptoms of criminal irresponsibility on the part of upright officials and technicians of the past, by assuring ourselves that these poor conscientious dupes were so blinded by superstition that they could not see the actual monster's evil through the blaze of inherited and confirmed divine right. Scientifically educated technicians, we felt, must by their training be immune to such superstition.

Yet to-day, in every despot-ruled country, technicians trained in the latest disciplines and discoveries of the physical sciences, are putting these resources at the despot's disposal. It is precisely because technical knowledge in physical science has advanced so far that men who are as ruthless as Genghis and as megalomaniac as Nero can continually threaten civilization with the destruction beside which the campaigns of the Tartar, horrible as they were, were little more than restricted military manœuvres and the burning of Rome hardly went beyond a cruel practical joke. Far from their training in the physical sciences giving the contemporary technicians immunity from superstition, and the intellectual insight and moral power to deny to the tyrant their essential aid (and so instantly to immobilize him) all that that training has done is to make the tyrant at last able to fulfil his homicidal dream, to turn all his subjects into inexhaustibly efficient assailants and the world into a shambles. So the cult of efficiency and doing one's duty without understanding whither the whole social drive is being directed is that which at length is able to destroy mankind. Lacking Right Knowledge, leading directly to Right Livelihood, all the succeeding secondary virtues can easily become only the alloys whereby soft vice and vague malignant dreams of hate become the most real forces in the world. The efficient technician, the conscientious expert, not looking beyond the well-doing of his special task, is only the third rank in an order which needs, if man is to be safe from self-destruction, two higher ranks which act as brain and eyes to these hands. Lacking these upper ranks of understanding and vision, mere technical integrity leads to the present nightmare, that whereas the majority of individuals are kindly, the States which they compose, sustain and love, behave like diabolical monsters. Bravely doing one's duty can lead very easily to dropping poison gas on children with a good conscience, because one is faithfully obeying orders. Such a conscience is only painless because it is dead. Yet there is no way out of this acute moral difficulty, this hideous stultification and miscarriage of simple brave conscientious goodness, unless Right Livelihood is seen as an essential step to true morality, and so to enlightenment.

Yet Gautama's solution of Right Livelihood by the begging bowl, we have seen cannot be ours. Even if the Western world were ready to keep mendicant monks, they should not accept such keep. For so doing they become accessories to a system which produces their food by denying their principles.

Nor is that all. Gautama accepted as proved that life had reached the highest limits of consciousness that it could attain in the body. He who would go further must abandon psycho-physical development and go on through psychical growth solely. Every monk was intending to achieve Liberation, the Nirvana which guaranteed, whatever else it refused to define, the assurance that never again would that individual consciousness find itself incarnate. There was the carnally-minded man who fed the monk. The only stage above such a man caught in illusion was the monk who had abandoned life and only looked for a higher consciousness which would mean his physiological end. We cannot be so sure. We cannot assert the impossibility of further psycho-physical evolution (of an enlightened spirit of enlarged consciousness still being able to develop in the body), until we have seen what type of mind is possible to men who are living co-operatively in an organic society which functions through a direct experience of the Whole.

When the trying out of such a community is seen to be necessary, many will urge that this is impossible. Such organizations have always failed. The whole tendency of civilization is away from such intentional societies. The development of the monastic orders themselves was into ever looser, less distinctive organizations, away from the "City-State," the complete economy of early Benedictinism, to the boarding house, the "bed and breakfast residence" of such Counter-Reformation orders as the Jesuits and the Oratorians. In their turn the Colleges have relaxed, the don becoming less and less distinguishable from any other professional or business man. The Colleges are coming to an end as contemporary social patterns, and are only being preserved with their Gothic and their gowns because of their picturesque archaism. If there is a contemporary vernacular social pattern it is the revolutionary "Cell," whether Communist or Totalitarian. True, during the nineteenth century, there were attempted, especially in those new states, where vague hope still filled the vacant spaces occupied in more settled countries by experience, many social experiments. Nearly all of these, however, rapidly failed. Indeed, the only ones which attained to *any* stability, did so because they depended on the leadership of a dominant personality. So modern America, in this respect, only confirms Father Tyrrell's sad judgment on medieval Europe: "No Order ought to outlive its founder."

This judgment on the history of these attempts is not, however, true. Some communities whose "viability" would have been considered hopeless—as for example the Mormons—have flourished and "made the wilderness blossom as the rose." And, even were it true that all efforts at intentional living have failed and in the past had to fail, that would not rule that to-day such efforts are not required and, if attempted, must collapse. Our civilization is not pursuing that course to easy affluence, to "permanent prosperity" which made men up to a generation ago believe that we could relax all social ties, and the looser we were, the less there was any general co-ordinating purpose, the richer and more creative we all should be. We need to know our collective purpose or we shall be driven as slaves to obey the ambition of some tyrant. Whatever returns, it will not be *laissez faire*.

Nor need our efforts fail at achieving such a corporate pattern and purpose, such a way of life which may make straight channels for the stream of evolutionary purpose. Such a society has never yet been tried. In nearly every case both the economic and the psychological methods which avowed intentional orders have used, have been very rudimentary and inefficient. It is surprising that they worked even for a year. We are, therefore, compelled to try out the experiment of a community existing expressly and efficiently to sustain its members so that they may attain enlarged consciousness, a community using all our social, economic and psychological knowledge to create its means and envisage its end.

Buddhism, Hinduism and Christianity have all had masters of consciousness who have attained complete freedom and insight while still in the body, and who believed that their system could be conveyed for us to follow, if we chose to adopt the necessary life. As has been mentioned above, the state is definitely recognized in Buddhist and Vedantic literature under the respective names of Samdithakkam Nibbanam and Jivan Mukti. Among the Christian masters Eckhart declares that he has attained to such complete enlightenment in the body that he no longer has any desire "to depart hence, which is far better." He writes: "I would gladly remain here until the Last Day." He stresses: "Many people imagine *here* to have creaturely being, and divine being *yonder*. That is not so. By that many people are deceived. I have indeed said that a man beholds God in this life in the same perfection, and is blessed in exactly the same way, as in the After Life." "After this life the potentially and partially realized shall be transformed into full reality. This transformation will not, however, perfect the experience of blessedness more than it is now."

If these pioneers could attain full consciousness while still in the body, the thing is not *ipso facto* impossible; psycho-physical evolution could go on, man as a social being could attain to a radical advance in awareness, the progress of humanity on earth could be extended. Granted that the men who up to the present have achieved superconsciousness in this life were psychological geniuses, we must also take into account on the other hand two things in our favour. Firstly, many of them were using methods of releasing consciousness which were rudimentary. Eckhart certainly had to transcend anthropomorphism and discard a thick crust of intellectualistic theology before he could achieve his freedom. The anxiety which he certainly suffered when his fellow churchmen broke with him must have handicapped, at least for a time, his attainment of complete detachment from his Church and his theology. Nor is there any evidence to show that he ever made his psychological techniques explicit, even to himself.

Secondly, we are *older.* Self-consciousness has grown, whether for better or worse. For worse if we do nothing about it. Because evolution is now psychical it is now so accelerated that we do, in fact, belong to an older section of the species than did these men, and it is, therefore, more urgent for the race that we make the radical departure, this psychic mutation. The time remaining during which our species may remain flexible and free enough to make radical advances in consciousness is growing rapidly shorter. Already we are showing many grave symptoms of degenerative specialization, loss of creative initiative toward our entire circumstances.

Alternatively, the fact that we come later in these last rapid stages of evolution can be for the better if we determine to do something. We are more self-conscious, our psychical development is at a more critical phase. If then we act, our action can hope to be more of a science than (as was theirs of the past) an art, or even a happy accident. Further, because our action may be scientific, that means that our process and technique may be comprehensible to any who would care to train. We must then give the world a demonstration of this manifest social pattern of avowed intentional living. We must show humanity a life-scale, working model of a complete purposive society, a society which has a sound and complete economy because it is based on a sound and complete psychology, a community which has all adequate means for full psycho-physical living because it is framed and directed toward the one completely satisfying end.

CHAPTER XIII

THE PATTERN FOR PROFICIENTS

THIS, of course, is an experiment. We cannot say what the outcome will be. We cannot say that we can achieve this psycho-social model as the vehicle of further psycho-physical progress. It may be that such a society could not endure but would volatilize, not because of any inherent fault in its construction. On the contrary, its very efficiency, the fact that it was constructed to make men able to achieve advanced consciousness might bring to an end our present arrested balance, where we remain transitional creatures tied to the present, and looking back to the past because we can descry no adequate future. Let the mind have true experience of that distortion we call Time; let the whole mind perceive clearly its goal and it might not stay, for it might realize that it had already attained its purpose, the reason for becoming incarnate. The Vedantists say that when a man knows who he really is, he ceases to need any longer a body. For that reason Ramakrishnan would not tell his great disciple, Vivekananda, who he, Vivekananda actually was. Full psychical release may so draw all the energies into one advanced focus that the psycho-physical combination may break, the body be dissolved or (perhaps a juster simile) radiated away. The purpose of evolution may be achieved.

Here it is worth while quoting the opinion of one of the world's subtlest psychologists, the great Hindu thinker, Sankara. According to him the very fact of incarnation and embodiment is a symptom that craving and fear still remain, for the body is nothing but a material expression of such states of mind. "All that we are is the result of what we have thought" (thought being, of course, all mental activity). When, therefore, we attain to the super-consciousness at which shrinking and craving no longer are felt, the ties of the body are undone, the blind will-to-live sees that it lives by consciousness and not by physical survival. We may last on for some years after radical understanding has been attained, because it takes time for the release to work itself out. Every thought took some time before it resulted in action, every action before it resulted in function, every function before it resulted in organ. The recapitulation, the unwinding of the coil, must also take time. So a man may, according to Sankara, be already in his thought, in his action and even in his function free, but the final core, or organ, remains still for a while unreduced. The echo still sounds after the voice is silent; or as the Hindus express it, the wheel still spins of its own momentum for a while after the foot of the potter has ceased to drive it on. Fail quickly it must, however. The man who has really experienced super-consciousness and made that experience native to him, is done with greed and fear, the two forces which precipitate the individual body.

It must be repeated we do not know whether the judgment applies in our case. It may not. The India in which Sankara lived and thought was a very different world from ours. Such a remark may be dismissed as being prompted by the vulgar delusion of physical progress. It is necessary, therefore, to repeat that everyone now living is a later special type than those in Sankara's world. He looked out on a society where individual self-consciousness was rare, a world of a few daring departures and pioneers and of vast numbers of stabilized routineers. Neurosis was low, society closely holding in an organic pattern all its members except the loftiest, and even their place was recognized. They were not escapists but "seconded" to be society's psychic pioneers. To-day society is in pieces, and though few as yet can find the answer, all ask what is life for and many if told could understand. The notion of evolution, of the flux, is understood more widely than ever before. The assumption that evolution meant inevitable physical progress has been disproved. The fact of evolution still remains. Show people its actual path, from physical advance to psychical advance, and they can understand.

We may say, then, that had Sankara been living in a society as broken out of its matrix as ours to-day, and had he been able to frame the intentional socio-economic society which now is within our power to frame, he might have decided that further psycho-physical development was possible. Granted that the final development must be purely psychical, purely transcending the material, still, some further steps may be possible and advisable before that final step. Buddha, and indeed all practical mystics, have taught that this present psycho-physical life is peculiarly favourable to a certain stage of spiritual advance. It may be that if we could make a psychologically intentional society or community, we might create a social organism in which the life of the spirit could be more favourably advanced than ever before.

We must come back then to that truly modern and scientific decision: only experiment can decide. We must frame such a society and whether it lasts or if it volatilizes, in either case we shall have carried out an essential test, an advance of essential knowledge and so have served the evolution of consciousness.

That question having been given provisional answer, a second more general one arises: granted that such a model society works, and that a few people who care wholly for the advancement of consciousness will be able to construct and sustain such a society, a society which will be the next step in human civilization, the social mutation which will carry humanity beyond the present deadlock of international anarchy, what then? Will large numbers be impressed with the fact that a way through has been found? Will they realize that a sane economic system is possible and only possible by a psychological training, itself only possible because the purpose of life has been discovered?

Again we cannot say. What seems probable is that in this discovery of the path of further evolution and in planning for that way of life, individual and social, which may follow that path, there would be constructed a further contemporary development of that avowed intentional living, that planned and rationed way of life which has successively developed in the companies of priests and prophets, the Essene and Hermetic societies, the monastic orders and—last of the term—the Collegiate Systems. In fact, there would be developed, as an answer to modem anarchy, a modern Order, which in the vernacular of our contemporary lives would serve as its predecessors served. It would have three functions: it would reintegrate the individual, for as long as he is psychically dislocated he cannot compose a sane society; it would rebase a society of unlimited liability on a sane subsistence through its own direct production of its economic requirements; it would preserve the learning and culture which is disappearing in the outer world. It would also (because it is a contemporary manifestation and therefore looks to the future, not to the past) deliberately forward research: economic, physical, but preeminently psychological.

This speculation in probability may, however, be easily falsified. Not only does history not repeat itself but many developments are cut off before they have finished producing all their possibilities. The sailing ship was still being evolved when the steamship sprang up, bringing to an end the earlier line. Civilization may destroy not only itself but those advanced variants which would preserve its essential treasures. The early Mohammedan invaders aimed particularly at exterminating the Christian monasteries, and later, when Mohammedanism penetrated into India, this particular hatred of the monk dealt, some historians think, the final blow to Buddhism in the land of its birth.

No Totalitarian State would sanction an order of New Friends within its borders, but would hunt out any such settlement and destroy it. It is always the psychical pioneers who are rightly the aim of Tyranny's most extirpating attack. The Christians were seemingly harmless, yet woke the destructive energy of the Roman State, tolerant of all individual fanatics in religion. The Manicheans were persecuted in every country. Kemal Attaturk hanged the Sufi masters in Turkey. Ibn Saud of Arabia hangs all Sufis. Balbo, the Italian Governor of Libya, has ordered all dervishes, Sufis and Fakirs to be suppressed. The Soviet refuses to let Ouspensky's books enter Russia.

On the other hand, that very intensification of means of communication which to-day gives such inquisitorial power to the Dictators, permitting them to ferret out any who differ from them and denying to dissenters those refuges of distance in which the monastic orders grew, may also tell the world that a new path has opened, and too many may set out on the way for all to be destroyed. They may win through to give the world the demonstration which could convince mankind that there is another alternative to the choice—democratic futility or dictatorial frenzy—that there is a path between the horns of the dilemma, a life of means and no ends or a life of false ends which destroy both means and values.

We have, then, two chances out of three that an intentional order might prove to be an essential step in man's further social evolution. It is, therefore, worth attempting, for certainly our present civilization shows every symptom of its inability to solve its problems. Only a new social mutation seems capable of lifting us out of the rut in which we find ourselves and which all our unintegrated efforts serve only to dig more deeply; and that mutation can only spring from a mutation in ourselves, a really new view of the universe, a radically new insight of Reality. Any partial remedy is worse than useless. We have to advance on all fronts at once. An unbalanced initiative seems only to increase our instability. We cannot have a sane policy without a sane economics. Simply increase our contacts and we dislike each other more and can damage each other more fatally. In a world which is mechanically drawn together, the blind struggle for supplies can only become more acute, the blind dislike of the stranger more deadly. Yet we cannot have a sane economics unless we have a sanifying psychology. Class war is, we now have ample proof, no improvement on national war. Indeed, horizontal war has this additional horror over the old-fashioned vertical war that is a cosmopolitan civil war. Add to this that it is a religious war, war which sinks from being a duel to a massacre, because your enemies are "sinners against the Light," and we see our culture stalled, slipping back to a state of anarchy when the one wish is to exterminate any rival. All morality disappears. Mercy, justice, truth, in turn are extirpated in oneself. Fascists and Communists, nationalists and socialists, are not lower types than the duelling professional soldiers of the eighteenth century. Though they behave incomparably worse, that is due to the fact that they are more conscious, more aware of the only means which they can see for attaining those goals. They are later, and therefore driven by a more urgent sense that they have little time.

We must put a stop to the international anarchy which is killing civilization. To do that we must transform an international economy which at present results in the international anarchy. The international and national economy can only be changed through a new knowledge of our natures, for our societies are projections of ourselves. Nor can there be any doubt that the greed and fear, which drive the individual, must result in a society which is first competitive and then anarchic.

If we want to avoid the collapse of civilization we must then attempt a threefold simultaneous approach to our problem. The problem is quite clear: men, in attempting to reach their goals, many of which can be presented as harmless or even beneficial, can see no way of getting there except by the employment of means which are clearly fatal. The world of "hard inescapable fact," in which they see themselves, held by iron circumstance and inhuman necessity, dictates that they shall use such methods and be destroyed by them. There is no escaping this grotesque and deadly dilemma, if the universe is as the realist individual sees it. Such a psychology dictates its appropriate economy which imposes its inevitable policy.

But this world of hard, inescapable fact is just that selection and distortion of reality which a fissured and individualized consciousness must make from the sum of the whole. The solution of our difficulty lies deep, but is not unattainable. We must expand consciousness beyond the fatal cramp and limit which individualism imposes. Then we shall see a world in which value and fact are not in contradiction but are two aspects of one reality and where means do not frustrate the attainment of ends.

This expansion of consciousness is an idea as unfamiliar and as impossible to the practical man, and indeed to all individualists, as was Copernicus's idea to all the practical and common-sense minds of his day. We cannot imagine that the way out of our dilemma, that the escape through the net of outer events which seem to have us toiled, lies in ourselves, any more than Copernicus's scornful critics could understand that the motion of the apparently free-moving heavenly bodies could only be understood by realizing that their vast, apparently detached movement was due to the movement of the obviously firm and immovable earth. As has been said, we have to-day to make a Copernican revolution in psychology. This means not merely that we must understand that we happen to things rather than that things happen to us, but that until we have direct experience of the enlarged consciousness which transcends not merely individuality but Time, we see events under a completely stultifying misapprehension. Unless we have purchase on a reality underlying Time, we have no fulcrum by which we can make any real change in events. This is the truth behind that hard-worn judgment of disillusioned revolutionaries: *Plus ça change. . .*

Yet the alteration and expansion of consciousness is obviously of extreme difficulty. It is a new evolutionary mutation, and such are always dangerous. The vast number of all mutations are harmful, many are deadly, only a very few lead to a new step in the ladder of life. Life, like the Sphinx, suddenly asks a riddle of its creatures. Some remain silent and sink into living fossilhood, making no response to the challenge. Others make a reply, they throw up wild mutations, horns, extravagant spines and encrusting armour, but none of their replies are the only right one. There is no reason to suppose that this law of physical evolution does not extend to evolution when it enters its psychical stage. As we have seen but must repeat, it is comparatively easy to throw consciousness out of gear. It is extremely difficult to get it into a higher mesh. As we have to attempt this new step of consciousness ourselves (for, because it is an attempt in higher consciousness, it must be done consciously; it cannot take place as a matter of growth), we have to assist this change in every way. Because it is so fundamental a modification, it must affect every level of our being, and these lower levels may abort this development. Unless the entire psycho-physical being can be moved, any extensive departure by the uppermost levels of consciousness may simply wrench connections. The body may refuse to follow and yet the mind may fail to get free. Tripped and trapped, the whole may fall in complete confusion. We cannot then, hope to achieve this supremely difficult transition by anything less than a complete, confirmatory economy. Our way of livelihood must be consistent with our aim—the extension of consciousness, beyond the strangulating limitation of the ego.

If, however, we do achieve such an extension we shall have the only answer to which men would now listen. The facts we have reviewed, the acutely grave state of world civilization (the breakdown of the inherited patterns), and the power to make complete new efficient ways of living, give us considerable hope, for two reasons. In the first place, such centres of the new life could choose their sites, so as to avoid being across the path of the degenerate conflicts which must shake the world. They could be real autarchies, sustaining themselves completely in very small and secluded areas. They will have the geographical wisdom not to settle, as for example, the Essenes did by the Dead Sea in what looked secluded but was in reality on the track of the Roman campaigns and between the poles of two forces—world imperialism and fanatical nationalism. In the second place, and this is much the more important, the psychological knowledge of such groups might win for them tolerance. For not only would they have an advanced anthropological insight (in which we must remember the martyred members of a proscribed order are nearly always lacking), but they would also have something to give, a service to yield to the rulers and their subjects in the dying world states, a gift and a service which these people could appreciate. It has been often proved that the one man who can move not only untouched but valued when revolution is most anarchic, is the physician. Both sides need him and his power to heal puts him above the battle. We have seen that the world powers are full of sick men, and the sickest are at the top. Physically fit, their psychological sickness, their neurosis, their paranoia grows and must grow. They need, and they know that they need, any who can really "minister to a mind diseased: pluck from the memory a rooted sorrow." They are contemptuous of a new economy and intolerant of any policy save theirs. But as they are all active psychologists they understand their basic need—a deeper understanding of themselves—and cannot but need and even respect those who have a profounder and more working knowledge of the psyche than themselves. They despise and destroy the pleasure-loving, comfort-seeking democracies and the vagrant and uncertain mystics and the passionate revolutionaries and even the gentle apostles of world-wide good will. But the service and assistance of one who knows more about the mind than they do, and who can save them from what the closest surveillance of their secret police cannot, their self-haunting nature in the darkness and loneliness of the night—such service and assistance they could crave, and the new proficient need not deny. He comes not to destroy or to denounce, but to further psychological growth anywhere. He has understanding of and compassion for even the Dictator and his Chief of Secret Police. The new athletes need not always retreat. They could, perhaps, at decisive moments, take the initiative. It is worth noting that Genghis sent for a Taoist master and listened when the master rebuked him.

So we may have an opportunity to give a threefold demonstration of the truth of our evolutionary hypothesis, our reply to the question: "What is the meaning of life, and can we co-operate with its purpose?" We may reintegrate the individual, so making him a whole man in striking distinction from the fissured, distracted contemporary man. That is the psychological proof. We may reintegrate the group by basing it on a co-operative subsistantive economy, by making an unlimited liability company which can give steady and substantial support to its members because they give it the security of their willing service through their direct experience of psychic unity. That is the economic proof. We may reintegrate society as a whole by showing that all humanity could live in peace, in plenty and in hope, once men realize the truth about evolution: that this Life Force is now continuing its activity by working psychically in us today and that all society could take on an organic order, coterminous with mankind, if this was understood; that those who are the front and growing point of evolution are distinguishable by their dynamic freedom from the self; that all humanity grades itself in a psychical evolutionary order, obviously recognizable by the degree of intensity with which each level gives itself up to the attainment of supra-personal consciousness. That is the political proof.

CHAPTER XIV

THE INSTRUMENT AND ENGINE OF CIVILIZATION

IN THE last chapter it has been said: "If we want to prevent the collapse of civilization." In an earlier passage it has been asked: "Can psycho-physical evolution continue, or has humanity reached the point at which only psychical evolution is now possible?" Though society, civilization, or at least humanity might continue, marking time or fluctuating between anarchy and despotism, in convulsions or in paralysis—is all further progress only possible by discarding the body? The answer to that question was, "We do not know."

On these grounds, if on no higher, we are bound to experiment. On the basis that we now know the purpose of life—the further evolution of consciousness beyond individuality—we might quite probably work out a psychiatry which would salvage any individual, who chose to pay the price—the price in radical modification of his life that such a standpoint incurs. We might, quite possibly, out of such reintegrated individuals, find assembled a society which would have a new economic stability, because it had a new psychological insight and aim. Could we hope to go further? Can we hypothecate a civilization which had adopted such a view of life and which, therefore, arranged itself so that those who acted most fully on that knowledge were inevitably those who were most advanced and influential in its ranks? Here we face the question raised earlier: Is this hope only for the élite or could all find their place in it?

This could come to pass only if there were first given demonstration of such nuclear societies, such life-scale models of the new way of life, so that mankind could know that such an initiative had been taken and had succeeded—that the engine was assembled and generating steam and, as soon as they wished, the average people, sick of having no lead, or one that dragged them to destruction, could take their places, tack themselves on, and be drawn in the direction of true progress. Such a supposition is not impossible, for the new step in civilization depends on the free choice of men striving toward a perceived end. Mankind might, therefore, decide so to act. Men cannot be forced, and, if they do not want order and progress they certainly cannot be made to adopt these things, though they belong to their peace.

If, however, men are to decide to go forward they will only do so if they see before them the manifest social pattern of avowed intentional living. A great deal of modern reaction is due to the blind disgust at our society of mere means, driving men under the sense of frustration and futility back to anachronistic models of the life of ends. Give people the vision of a life of ends above and beyond (not below and behind) the life of individualistic liberalism and much reaction would be deprived of its appeal. Nor need we, nor should we, require that personally involved and critically-minded people leap out of their chains and ties and take on *instanter* the fully intentional life. They can only, and they must only, begin where they find themselves. Feeling their inner conflict, they may well wish to inquire into, if not to adopt, a training which manifestly cures this condition and leaves in its stead a whole individual. When they learn that the cure requires a radical modification of their whole lives, many will hesitate. As, however, much of their anxiety and inner conflict springs from, or is exacerbated by, the economic system under which they live, quite a number may be able to put two and two together and, seeing that their psycho-physical condition and their socio-economic system are interrelated and interacting, they may be prepared, for the sake of curing their distress, to enter on a psycho-economic way of living. This means that the foundation of some kind of Collegium is likely. We have seen that there is a long succession of patterns of avowed intentional ways of living, running like a thread through the history of our civilization and coming out most clearly when the traditional conventional ways of society have fallen into disorder. The monastery was the vehicle and expression of the essential tripartite society during the Dark and Middle Ages, and we see in its system the essential psychiatry expressed by the discipline and worship; the essential economy by the agriculture and craftsmanship; and the essential policy by the preservation of the classical scholarship and the development of the scholastic philosophy. The college-composite University, in turn, was the vehicle and expression of the Renaissance.[1]

The University college allowed the individual more freedom and privacy than did the monastery. The Chapel, the Library and the Halland-Common room were the three pivots round which the corporate life revolved. Scholarship, however, became the dominant exercise—as routine worship became the monastery's. The monasteries neglected their economic aspect and function. So, in turn, did the Colleges. The Colleges also neglected the psychiatric function. Learning became for them sufficient answer to life. The fact that the subjects of most of their studies were not co-ordinated to present the mind with a single picture of reality and of the scholar's place and function in that whole; the fact that intellectual understanding affects hardly at all the basic will, these grave facts the College overlooked.[2] In consequence, to-day it is treated with the contemptuous respect allowed to corpulent senility, the same respect the monasteries still lived upon in the generation before their dissolution.

That does not mean that society—especially when its conventions are as uncertain as they are to-day—can survive without consciously devised social patterns of avowed intentional living. Because the actual form, in which such intentional living answered men's needs in the Middle Ages and in the Renaissance, are now out of date and cling on merely as archæological curiosities, that does not mean that further development is not required in forms suited to our contemporary lives. On the contrary, the family and the club are seen to be

1. As the Christian monastic system began in Egypt and only after achieving its psychiatry and economy did it, with Benedictinism, enter its third completing phase, its policy, its peculiar contribution to the salvaging of civilization, so too Universitism began to construct its distinctive pattern of life, its treatment for the individual and the group, some generations before it produced its distinctive contribution to its age, the organizing of Humanism and the New Learning.
2. The intellectual is always in danger, even when a mystic, of making the mistake, common at least since the time of Socrates, of imagining that the intellectual recognition of an evil, will mean its abandonment and of a good, its achievement.

increasingly inadequate units for building a society capable of handling and expressing our present type of psyche, of economy and of policy. From the history of the past, we should expect that it is here and now, in answer to our topical confusion, that corporate living would develop its contemporary answer, an answer in the succession of the monastery and the college, those earlier answers to the confusions of the Dark and of the Middle Ages.

We may then, from our present need of such a form, and from our knowledge of past forms which met past needs, indicate in outline what features such a contemporary form might show. Blue prints are not in themselves costly things and may save considerable future expense. Some kind of unitary social organization will undoubtedly be precipitated by our present threefold tension. Our acute need for freedom and peace in ourselves, for proportioned and just distribution in our economy, for order, enterprise, and purpose in our society—any one of these is sufficient to cause a social crystallization. Unless, however, each of these three needs is equally represented in the new social pattern, it will start unbalanced, grow tyrannous and end in collapse. It is not possible for any social group to succeed in achieving a dynamic balance unless it clearly envisages its goal and understands the means through which that goal may be reached. Of Lenin's many tragic mistakes, few have proved to be more costly than his contempt for a clearly envisaged goal, which he dismissed as Utopianism, and his deliberate disregard of the inevitable connexion between means and ends when he dismissed justice and even truth as "bourgeois virtues."

Since the monastery became an anachronism and the college fossilized, there have been repeated efforts to design and to work contemporary intentional social systems. The first thing then is to learn, by the mistakes these systems made, what further features must be added to a planned organization if it is to endure. When we study these nineteenth-century experiments in Europe and America it is remarkable how the errors which proved fatal to them group themselves under those blindnesses from which the eightfold Buddhist way is designed to lead into understanding and liberation. The first mistake (which ranges from the co-operator Goodwin's with his futile little colony in the early nineteenth-century to the communist Lenin's with his gigantic vivisection of an empire in the early twentieth) is the failure to have a clearly stated purpose, a definite understanding of the material with which one is working, the human character. What are its capacities and what can be its aims? To fail here is to fail at the first step of the way—Right Knowledge. As Max Eastman has pointed out, all these nineteenth-century revolutionaries are pre-psychological. Obsessed with economics they make certain assumptions about human nature—that it only requires economic plenty to become morally perfect; that it is fundamentally united and only kept apart by anomalous individuals, the capitalists, who divide and rule. Then the revolutionaries proceed to blame, and if they come to power, to destroy, all who fail to act up to their assumptions. The Right Knowledge which gives us our aim is the discovery that we should and could continue evolution through the enlargement of consciousness, that this is the aim which all mankind is blindly attempting to achieve but that as humanity always includes a wide range in the degrees of consciousness, only those who are most advanced are fully conscious of the meaning of their blind striving.

The second mistake has been the lack of Right Will. If a new order is to arise and grow, it can only do so if, in addition to the clear understanding by its members of their aim they have also trained themselves so as to be able to make their conduct accord with that purpose. Now that we know how slowly the deeper directive and controlling will comes into line with what reason has planned, we have less excuse for overlooking the essential need of self-training. Now that we know the methods whereby that will can be approached and does respond, we have good reason to hope that if we so act, we shall get results where others had to fail. The ignorance of the subconscious mind, of its powers, its inaccessibility to ordinary argument, and the way in which it can be contacted, explain the community failures of those unaware of these facts and hold out hope to us who know. To an order which has a definite aim we must then add a novitiate. None of the ancient and medieval orders omitted this essential preparatory stage. We have seen in the Essene order the novitiate was a three-year preliminary training. The aim of a contemporary order is, we must repeat, threefold, and its three functions develop successively. First, it is a psychiatry. This must be so because our societies are projections of ourselves. A fissured psyche gives us a conflicting society. This is the only diagnosis which penetrates down to the source of our present anarchy. To introduce into an ordered and intentional way of living, as members, those who are still in the neurotic state of "normal" individualism is to poison the organism, just as a virulent bacillus kills a healthy body and perishes itself in the destruction it causes. When an individual has gone through his initial training, specifically directed to reduce the strangulated consciousness, the inflamed ego, then a healthy social organism will keep him healthy, as a healthy body controls and co-ordinates the life and action of all its cells. Until he is himself healthy—it must be repeated—all his efforts must be directed to attaining health. This is not psychical hypochondria. On the contrary it is a sane estimate of reality. It is not because the individual is self-centered that he attends first to himself, but because he wishes to get the self out of the way, in order that he may see clearly his place in a common task and because he realizes, with an active humility, utterly alien to the self-importance of the hypochondriac, that until he is "reduced" he would only spoil the lives and purpose of those higher types with whom he would co operate.

If people have the Right Understanding that civilization must have a new model, and the Right Will to give themselves to the building of such an order, they will welcome the opportunity to train for such a life and will certainly not grudge such time as that training may take. Those who have not reached Right Understanding, who have not discovered the purpose of Life, chafe at delay and grudge time. Even in socially-minded people this is all too common. It is, however, a fatal impatience, springing from the old mistaken view of life: that "You only live once," and that the only possible human aim is to achieve material well-being, in that little span, whether for oneself or for one's fellows.

The third grave mistake of modern model communities has been the lack of continuous training and psychical exercise in the admitted members. Right Meditation and Right Contemplation can never be pretermitted, any more than right physical diet and exercise. Meditation and Contemplation are the breathing exercises of a newly-born consciousness.

Unless these are sustained, this higher consciousness asphyxiates, and, like many forms of physical asphyxiation, death may steal upon us without our being aware of its approach. True, a few modern communities and some which have survived from earlier periods have maintained religious exercises, but these have often proved even worse than useless. In anthropomorphic prayer, requests addressed to an external magnified person too often merely relieve the individual from using even his surface will for inhibition of his ego-impulses while giving no control over the ego itself.[3] More important than its economic production, because more fundamental, is a community's psychical intake. Only when this is placed first and is adequate, will the ego be sublimated, will consciousness begin to expand, Right Livelihood be natural and productive, the Thought, Word and Deed of each constituent integrate with that of his fellows, and, so only, will social evolution continue and advance.

If, then, this is the way in which may be avoided the mistakes which have destroyed or rendered unfruitful modern attempts at a contemporary organic society, we may suggest the type of social organism the above rules would construct. The first thing that is clear is that such an organism would have three functions and so three aspects or ranks. We have seen that there would be primarily a novitiate, a training for those whose self-dissatisfaction had brought them to the stage at which they are aware of their fissured consciousness. If they seek help but are still so self-ignorant that they think the fault lies fundamentally in their circumstances, and that, therefore, life in a community radically different from that in which they find themselves would solve their problem, they must be shown that this cannot be so. When they have integrated themselves then they will be fit for, will benefit by and will benefit, a community of integrated fellow members, not before.

---

3. Lord Curzon, one of the most intellectual and capable of administrators, whose worth was almost wholly stultified by his childish inability to control a savage nature, once remarked with a tragic complacency: "Every morning I pray that I may be kindly, considerate and not lose my temper and every evening I find I have failed."

The postulants need not, however, feel that they are psychological paupers. Indeed, it would be highly undesirable for their own sakes that they should. That feeling is the tragic fate of too many who are psychoanalysed. If they are to progress and indeed integrate the psyche and aim at expanding consciousness they must look forward to taking, and intend to equip themselves so that they may take, an organic part in a living social organism. Therefore, as part of their occupational therapy, while undergoing integration, they can either serve the community as lay brothers, undirective, non-voting assistants who are bound by no rule at present except the full recovery of their psychical health, or they can, while still in the outer world, serve the community by informing others who may be willing to undertake the task of reintegrating themselves, and by making the world understand that an order exists whose first duty and task is to lessen the neurosis now epidemic throughout civilization. They would be the modern and progressive form of the Franciscan Tertiaries.

The second rank would necessarily be associated with the actual economic aspect of the community. Having already achieved the first stage, the psychiatry, the integration of the self, they would be ready to "come into residence," to become full members, joining in the integration of a society. They would run the centre at which such postulants and associates as wished to come for help in self-integration, might reside for treatment for long or short periods, and to which all associates could look as their base, the centre where they can always check up the progress they are making and take their bearings—not a retreat but an advanced outlook point. The life of full members would be predominantly economic, for an economy which springs from a sane and progressive psyche is a lofty life, a conscious intentional body politic. So living they would be constantly working out for humanity the life-size model of a contemporary purposive life, a life which demonstrates that all the psycho-physical needs and means which man requires may be obtained without conflict or competition, and, moreover, that such means may be ordered so as to assist and not thwart the further evolution of man, the attainment of the only end which can satisfy him, a consciousness expanded until it sees and co-operates with a supra-animal natural order and state of being.

Such a community centre, as it would have an actual location, may be briefly described. Local conditions must modify details: The essential skeletal structure of such a community would, however, be the same everywhere because there would have to be the same broad framework to express the community's essential functions and aims. In a phrase, the monastery was too close, the college too loose a structure. The college made the important discovery of the necessity for a certain space of privacy. Undoubtedly, privacy is not so vital a matter in a community of integrated individuals, as it is among members whose only tie is a love of learning. Such a tie is never of the strongest. It often became frayed when it had to hold many with such severely dislocated psyches that they could endure close association no better than inflamed joints can bear to be pressed on each other. Nevertheless, as our present consciousness, even when integrated, cannot be continuous but must reciprocate between sleep and waking, so the creative mind, the expanding psyche, requires to alternate between company and solitude. Indeed, the more stimulating and apt the companionship, the more need to assimilate, order and develop the nourishment and stimulant received. Even the thirsty cannot swallow without check. Each member would, therefore, have his own room for solitude and individual work. As has been said earlier in this book, meditation itself needs an alternation. Nearly everyone (except those advanced ones who can pass at will into contemplation) finds the best results when this work is done by a group, every member of which is also practising it in solitude. The one exercise enriches the other. A certain guaranteed degree of privacy is then necessary for even the highest activities of the community.

Beside this necessary retreat for each, the community would have its centres for all. The Collegiate System was too casual about the functioning of these essential organs. These are first the hall and combination-room where intellectual exchange takes place—that symposium which Graham Wallas saw as an essential creative element in Greek thought, lacking in our learning to-day, and indeed only able to function fully among a group which "uses the same language" and is aiming at the same elucidation. Next there is the library-laboratory—the place where intensive study and experiment can be carried on. It is necessary for a pioneer community to know what other pioneers are doing and have done, and to translate out of their traditional phraseology or inadequate expression into the community's vernacular. In many archaic forms valuable observations and rules lie disguised.

It is necessary also to be able to experiment. Each community, if it is to be complete, should number among its members men who not only have adopted the general discipline of the order, but who also have, severally, the particular equipments of a psychiatrist, a physiologist, an endocrinologist, a dietician and a biochemist. Problems which concern all these special knowledges must arise in any community which is endeavouring to reintegrate individuals and, from such reintegrants, to construct a society whose end is the constant extension of consciousness. We do not know what strains have to be taken by a physique which is sustaining an advanced psyche. Intentional evolution must be a dangerous adventure. As the late Dr. Gamble wrote when discussing the physiological problem which even our present moderate and insufficient degree of consciousness entails: "Already the lines of communication between the upper centres and the lower are so extensive that trouble is bound often to arise from this cause solely." Orders are continually being sent down by the fore-consciousness requiring demands upon the body which can only be met by the nervous system imposing strains which the body cannot endure. Ignorant of our nature, our individualized self-consciousness cut off by the limen from any direct knowledge of our complete psycho-physical being, we, in our impatience and greed, gradually bankrupt our organism, which, first, becomes functionally deranged and then organically damaged, when we have grave physical lesions, such as internal ulcer, diabetes and many other diseases whose immediate cause is due to deranged function, itself due to profound conflict between the surface consciousness and the deeper controls.

Granted that an integrated consciousness would remove these "kinks" and strangulations which cause many of the present breakdowns, on the other hand, the still further extension of consciousness, if it is to continue in the body, will require careful watching and such watching will be repaid with essential knowledge. The mystics record that great physical pain attends certain shifts and enlargement of consciousness and though this may partly be due to psycho-physical ignorance, most who have practised carefully certain Yogic exercises, which aim at the same thing, note that they do experience certain pain and that such pain is considered to be an unavoidable symptom of this change. The charge of basic, evolutionary energy, though being released into higher consciousness and to develop higher faculty, is being led off so quickly that pain is experienced.

To return to the structure of such a college, most important of the three essential centres of the community is the centre for group meditation. Even in the austerest, least expressed, and most intuitive of all forms of Buddhism, Zen, the meditation-hall is an essential organ of the system, indeed it is the very heart and brain of it all.[4] In the Zen organizations there is a constant reciprocation between such intensive group meditation and the occupational therapy of garden and kitchen-garden work. This, undoubtedly, makes a sound system for any intentional and purposive order. The body's need for exercise, the nervous system's need of expression, can best be met by integrating them with the production of economic requirements. An efficient, intensive garden and farm would complete the features of an intentional social order.

A third rank has, however, been mentioned. What is its nature and where is its situation in this plan? As the associate member comes under the class of those who are predominantly benefiting by the psychiatry—the reintegration of the individual psyche—and the full member comes under the economic class whose main activity is constructing a new economy based on a direct sense of a living group—a real body politic—there must be another class above the full group members, as these are ahead of the associates. Without them the New Order would lack its fruits, the new organism would

---

4. cf. Hlow Philo's Therapeutæ, although living with the utmost austerity, kept one room in their simple dwelling solely for their psychological work.

have hands and ears but not eyes.

We may use again the analogy of the monastery and the Collegiate System (an analogy which may well prove to be more than an analogy, may be found to be a true homology). In the monastery there is the novice, and there are the fully-professed members. But there are also officers such as sub-Prior, Prior and Abbot, and there are less defined but more psychologically real ranks in those proficients who are permitted to undertake advanced and exceptional exercises because it is recognized that they have attained to peculiar athleticism of spirit and singleness of purpose. The natural emergence of a third advanced rank, above the two normal classes of learners and educated, is even more distinctly evident in the Collegiate organism. Here we find the undergraduate learner, and the graduate who has learnt and so is qualified to become a Fellow of the College, one who has attained to equality with educated men. But above these we find a third distinctive type, the Doctor. This type is qualified to teach, and, as the English word recognizes, by associating the word and its rank mainly with medicine, to heal also.

It is this third type which must emerge from the new order. There is another social analogy by which we may describe this third rank of Doctor Proficients—that is the analogy drawn from the ancient Laws of Manu. The Doctor Proficient will be the modern form of the Brahmin. The Brahmin suggests notions of caste. The evil of caste is, however, its rigidity, not its organization. Liberal, individualized Democracy is failing, failing politically against the Dictators, failing economically to solve its own internal stresses, from lack of organization. The disappointment over the limitations and failures of economic planning should show us that what we need before all such attempts is the laying of a foundation on which they may rest. Before economic planning we must have psychological planning. The Democracies have an increasing, suffocating and strangling amount of false organization, of bureaucracy and its red tape. What they lack, and are dying of the lack of, is a real organic psychological relationship between all their constituents. Then the State would not be something alien and Government something to be eluded or captured, but at last it would be possible to say: *L'état c'est nous.*

Caste has no evil but only good if its two original and basic rules are observed: (1) that anyone who can qualify for the higher or highest ranks may and must be admitted; (2) that the higher the rank the less has the individual any amusements or private possessions, while in the highest rank any office or direct power and influence must be ruled out. These large social implications of the word can for the moment be left aside as also the accompanying question; if the third and highest rank of the new order is the Brahmin type, must not, then, the type next below, the Fellow of the College, be equated with the Kshatriyas, the executive (ruler-warrior type) to-day, and the subject-in-training with the Vaisyas (the merchant type), to-day the technical official and technician? And that being so, where comes in the general basic rank of the unorganized (or unorganicized) masses—the Sudra routineers?

Before we deal with this development and how the Neo-Brahmin might bring it about, we must consider, as exactly as we may, what sort of man he would be, his capacities and powers. That is dealt with in the following chapter.

CHAPTER XV

THE NEO-BRAHMIN

WE HAVE seen that the first and simplest type of those who feel they must adopt intentional purposive living are patients. They feel only that they must cease to be the pawns and victims of their own moods and misapprehended nature, that they must win their liberty from internal accident and so achieve the power to endure external accident; these, the lowest of the three ranks in the new order, are specifically concerned with the psychiatric problem, the reintegration of their own fissured psyche.

The second type, having attained interior reintegration—and with it the power to endure external accident "with an equal mind" (the limit of the Stoic's ambition)—can and does take the initiative against circumstances. He can live in an intensely co-operative relationship, the only real communism, with his similarly educated Fellows. He can achieve the only planned and purposive economy, where means are ample, controlled, and perfectly adapted and directed to make possible the attainment of clear, satisfying ends. He can very largely disregard accident.

The third type could be as beyond and above this level, as this level is beyond and above the level of those beginners, who can only be concerned with their own condition and with their recovery from it. Such a consideration by itself certainly suggests that this third Neo-Brahmin type would of necessity be concerned with the world. He has overcome his own internal conflict and completely reduced his own psychic fissure. He has rewon the initiative toward life and circumstances so that he no longer needs the distraction of amusements nor is he the prey of any anxiety from fear of accident. He is even beyond the state of the Mastership, the Fellowship which builds up an economy, a communist collegium so that there may be an economic independence of an economically cataclysmic civilization, and within this advanced and isolated organization the good life, impossible without, may be achieved. It would seem that the Neo-Brahmin cannot remain content with a "fugitive and cloistered virtue." In his emergence we see a hope for mankind as a whole. It is here then, that we can come back to the main question with which the last chapter began. Would it be possible, not merely to achieve a new order of men devoted in total independence of civilization, to winning the next step in evolution, higher consciousness, but could this order take the leadership of mankind and help all to escape cataclysm?

Up to the present we have been confined to considering a very restricted solution of humanity's problem, and we need constantly to remind ourselves that lacking the third type, this Neo-Brahmin, the best we can hope for is the salvaging of a few from the general wreck. We cannot honestly hope for anything more. This phenomenal world is a projection of ourselves, of our psyche; this civilization is a projection of our sundered fore-consciousness, our strangulated individuality. It is pre-psychological to expect any real change for the better unless we can change ourselves, alter our human nature. It is not pessimism but scientific realism which compels any modern mind to face the fact that any social solution short of a psychological solution (in the profoundest sense of the word) belongs to a hopelessly misapprehending, subjective, "pre-Copernican" outlook. Therefore, all that could be suggested even on the most optimistic expectation was that a few would surmount the present difficulty. Psycho-physical evolution might go on, the pent forces which are destroying us might find their right and sufficient expression, we might apprehend the true meaning of our evolution, an expanding consciousness might continue to develop in a healthy body—but only in and through very small psycho-economic units. In fact, lacking the third type, the Neo-Brahmin, the two subsidiary types could only create another stage of monasticism to answer another, and perhaps the Ultimate, Dark Age. By far the greater part would perish: "a remnant shall be saved."

With such a power as such recluses would command, only the power of a costly-won example of sane living, the drive of the world to destruction under the control of the daimonic power-types, which now hypnotize it, could not be arrested or deflected. Even the easy-going, still-tolerant defeatist does not believe in the integrity of that example. He is sure that these organized intentional livers are even more defeatist than himself. He has his Freudian explanation pat and ready. The mere fact that these poor escapists have not even the guts to "eat, drink and be merry for tomorrow we die," proves that they are making even a poorer reaction to disaster than he. No, he is not likely to be won over by their example to a life of psychological training and exploration. He is safe from any such appeal in the assurance that he has explained away this ascetic action as a morbid and cowardly retreat. Any simple devotion and self-sacrifice no longer awakes even his ineffective admiration or respect. He dismisses what his parents would have grudgingly reverenced, with easy contempt as "mere masochism."

If there is to be any general escape from the oncoming collapse, the masses of humanity—whether they consider themselves educated or not—must be confronted with something much more impressive than a few out-of-the-way examples of a new sort of college where people who cannot stand the current strain have themselves reintegrated, and where those who splice them, live a meagrely successful economic life with a vague Quaker background.

We see, then, that only the rise of a Neo-Brahmin type could be an adequate answer to humanity's case. The condition is now so advanced and grave that it can only yield to heroic measures. In the Mahayana tradition of Buddhism it is said that whenever the world grows so evil that even the kindly can see no means whereby men of power, authority and action can abstain from cruelty, treachery and violence, then it is that a Bodhisattva, one who has attained to the possibility of complete supra-individual, extra-temporal consciousness, but who abstains from this private deliverance, reincarnates himself. For only a spirit proportionately lofty and powerful as humanity's temporary masters are low and brutal, can serve in such a pass. Such a mythos, such a religious fantasy may seem to have no bearing on our present problem. Our task, our pressing issue is to save civilization. Transcendental talk must sound completely beside the point. Yet we can all now agree at least on one point: that all ordinary and accepted measures for remedying our condition or even for arresting collapse have failed and we may add, that the forces which are causing that collapse are not economic forces. They are unprecedented morbid psychological energies which we find it almost impossible to diagnose, still less to prescribe against. It is, therefore, surely not too fantastic to suggest that psychological diseases shall be psychologically treated.

Nor need we take a leap into magic and demand a miracle before we will move a step or consider a cure. Seen from one angle our present chaos and peril has nothing unexpected about it. It is not a bolt out of the blue, a causeless cataclysm, any more than a waterfall in a river's course is a cataclysm. The persistent behaviour of all statesmanship, the complete lack of any ordinary plain dealing and honest conduct, still less of any anthropological or psychological insight in international affairs must have led to such disasters as we at present endure and must foresee. Our psychology, long arrested, had long resulted in chronic evils. But the spread of these evils, from local epidemics into pandemics, which could threaten civilization's existence, was confined. We lacked the physical powers to express our unrestrained greed, fear and malice against those strangers with whom we felt no spontaneous restraints. We had the physical powers to abuse our control over our neighbours. This we did: but, even in this restricted field, our violence, the full mischief which our local, short-ranged powers permitted us, was often restrained by a natural shrinking from hurting irremedially men and women whom we know, who are like ourselves, and whom, when our spite is past and our temper mollified, we shall see, their damage reminding us of our subhuman behaviour. Now our physical powers have advanced to world-embracing and world-destroying proportions, while our psychological powers have remained arrested growths, have actually atrophied.

Any sociologist ought to have been able to calculate that if physics and chemistry continued to advance as applied powers at the rate of progress pursued throughout the nineteenth century, and psychology and ethics remained arrested and increasingly confined as academic abstract studies, then civilization must capsize, as soon as physical power was adequate to destroy physical resources. That date has arrived and could have been forecast as clearly as a waterfall can be foreseen by a map reader following a river bed.

That is the negative side of the diagnosis. A morality as low as that currently practised by statesmen for centuries past was not sufficient to save the world from chaos. On the contrary, it would have flung the world over the brink long ago had these men's powers been equal to their will. We were saved, not by our psychological good sense but by our physical stupidity. Such a consideration, though not cheering, at least is sobering and should save us from panic. We are not faced with devils to-day, but only with men capable of doing what our elder and honoured statesmen would have done had they been able and empowered. The present leaders may be daimonic, sleep-walking fulfillers of destiny. They are not diabolic, loving cruelty for itself. Not yet. We are not confronted with any cataclysmic decline in morality, but only with our century-long arrested and inadequate psychology, suddenly required to balance accelerating physics. We have to produce, if we are to be saved, a psychology as practical and powerful as our physics. Otherwise our destruction, though as natural and as slowly matured as a typhoon, will be as rapid.

The positive side of the diagnosis is certainly not without hope. It is that whenever in the past just, generous, and courageous dealing has been used, it has been effective in a way so much surpassing the successes of trickery and violence that the actual records of it seem almost miraculous. Such examples and the technique which underlies them can be studied in Case's *Non-Violent Coercion* and Gregg's *Power of Non-Violence.* There does not, in most of the examples, seem to be involved any superhuman or "paranormal" capacity. On the contrary these nearly all seem to depend upon something which may be called an animal capacity. As all animal tamers have now discovered, fear has a smell (see "The Scent of Fear," *Reader's Digest,* January, 1938, p. 28) probably due to the same mechanism which lays back the ears, raises the skin on the scalp and back, and bares the teeth. Most likely this is a spontaneous reaction provoked by the suprarenal glands which respond instantly to any powerful emotion and which release into the blood their secretion, adrenalin, as a preliminary aid to conflict. The adrenalin is quickly carried out to the functioning sweat glands. The being, animal or man, who has roused one's fear, smells the adrenalin and so warned that one is hostile and physically (if not consciously) ready to fight, itself experiences a spontaneous impulse to attack. Arrest this vicious circle of interacting reactions and the mounting anger subsides. It takes two to make a quarrel and the necessary field for an explosion cannot be built up if one side will not play its part in increasing the volume of resonance. This, however, is no easy solution and the pacifism which wishes to believe that mere avoidance of aggression will assure peace, is mistaken. The mistake largely arises from the peace-loving being sated and the trouble-makers being the hungry. Too many pacifists are not prepared to cease to benefit by the wrongdoing and trouble-making of theft grandparents and, even, of their parents. They forget that the wronged must have longer memories than the wrong-doers.

Yet pacifism's mistake goes even deeper than the attempt to avoid restitution before asking for a general amnesty. The savage dog attacks regardless of the fact that you may have paid his master's bills. Why is that? It is due to two factors: to memory and to being reminded. The dog, partly through its heredity, its race-memory, and partly through it individual memory, its brutal training, has a strong recollection of violence and little power to restrain its own suprarenal excitement when these roused glands come into play. For all the carnivores have large suprarenals and comparatively small thyroids. So the "glands of conflict" tend to over-master "the gland of reflection." A hopeful consideration at this glandular level, is that in man this balance is reversed, the thyroid being comparatively large against the suprarenals. To the savage dog, always "on edge," the appearance of any stranger is a provocation. He has to utter an hysteric challenge to reassure himself, and such snarling, growling and barking naturally awakes nervousness in the stranger. The train is fired. The point at which it can be damped and extinguished is, of course, where the stranger sees and hears the dog snarl. He can inhibit his own fear reaction. If this can be completely controlled the dog, which is sinking to a very simple level of reaction, is thrown out. The initiative now lies with the man and if he can speak and behave reassuringly to the animal it must either respond or give him a wide berth. He is, evidently, at the least, someone who can· not be attacked.

But the man's natural, spontaneous reaction of fear cannot be consciously controlled. The surface stoical will can make the man stand rooted, can control the large muscles, but unfortunately all this surface control only makes matters worse. For not only does such inhibition and repression not control the suprarenals, which are doing all the mischief by rousing the dog to increasing rage, it actually makes the discharge of adrenalin increase. The high nervous and muscular tension calls for more stimulant and the more the man sets his jaw and his whole body becomes tonic, the more he reeks of provocation to the hysterical dog. The surface mind always aggravates any psycho-physical distress which it strives to allay.

Such is the pacifist problem put in its present frame, when suspicion and wrong, long left to fester, have led to hysteria on the side of the wronged, and panic concessions on the side of the still-possessing. Civilization to-day has reached that last-but-one act of the tragedy which begins when a nervous provocative man falls in with a dog which he has made savage by beating and privation. In act one they meet, and the man, who still feels sufficiently assured that he has the whip hand, drives the dog away. In act two the dog suddenly confronts the man in a place where he cannot use his whip. In act three (which to-day we have reached) the man in a panic throws down his whip, throws some of the contents of his full pockets to the dog, and takes to his heels. Act four may terminate the play by the dog going after the whipless man, pulling him down and worrying him to death. Or there may be five acts to the play, a fourth while the man is let run a little while, keeping the dog at a distance by throwing him all the food he has on him (and he yet has much) and a fifth where, all his goods exhausted, the man himself falls a prey to the beast he provoked.

There does not seem any other conclusion to the play: terminated in four acts or extended to five, it must still have the same ending.

To deal with such a crisis amiable good will and timid non-provocativeness are not enough. The provocation has been given, the man no longer stands as a harmless figure in the dog's mind but as a monster to be destroyed if ever the dog is to have an endurable life. An immense initiative needed if such a situation is ever to be redeemed from what is now an almost automatic degeneration and an inevitable cataclysm. Restitution now comes too late. As in the psych logically true mythos of the sibylline books: the price, which would formerly have bought complete understanding, no will not purchase even a beginning. The time is past when ordinary good will could have been understood as generosity and not immediately assumed to be resentful weakness conceding what it cannot hold in order to gain time to recover its power to retake the offensive.

Nor is mere weakness, the will to yield all, any more sufficient. The mere sight of a creature running away makes any animal pursue. Fear, as we have seen, is just as provocative as anger and even more immediately dangerous, for anger may temporarily win, while fear must always and instantly lose. The mistranslation "The meek shall inherit the earth," has led to much misery and done much to delay man's social evolution to a dangerously late moment in time. Praos rendered as "meek" is given in the French New Testament as *debonair,* the very opposite of that shrinking, negative, provocative attitude which we understand as meek; a certain gay self-forgetfulness and a spontaneous wish to put others at their ease. Yet the actual Greek word has a deeper, more intentional connotation; more adequate to our crisis than any cheerful good breeding however gallant and generous. Praos is the "trained" spirit, the man who is the deliberate, powerful and adequate opposite to the wild, untamed creature, man or beast. The trained do inherit the earth, and no one else. The trained phalanx pushed the packs of Persian levies out of history. The trained legion thrust all the battle-loving undisciplined barbarians before it, until the governmental will which operated the legion, failed. Training at this low level availed, until, failing to develop beyond the rudimentary discipline of drill, failing (with the Stoic's bankruptcy) to have a correspondingly advanced and adequate training for the controllers, the Imperial System fell into atavistic uncoordination. So we have to-day the disciplined technicians, having crushed barbarism but lacking among their leaders any adequate universal construction to put in its place, turning on each other, supplying atavistic leaders with the power which the barbarians lacked, the power to destroy not merely the bodies but the thought and frame of mankind, his records, his achievements, his corporate traditions and associations, the moulds and armatures of his intelligence and feeling.

Yet this training is neither impossible nor a leap in the dark. It does develop inevitably and step by step out of that ordinary and spontaneous considerateness, that generous dealing, that fine behaviour of the debonair which every human being likes, in the moments when he is aware of others, to imagine that he possesses and that they recognize in him. In the preliminary stages of human contact before misunderstandings and sudden resentments have worsened the relationship, common justice and patience will resolve difficulties which to the suspicious and those who start with the false assumptions of individualism, seem insuperable.

The classical example of the success of this procedure, this level of trained good will, is the founding of Pennsylvania. The detailed descriptions of Penn's famous treaty with the Indians when taking over what they might well have refused to yield, brings out three basic factors in this typical, non-violent deal. There were present a philosophic, a psychological and then finally a legal agreement. With true anthropological insight Penn found a general premise and basis between himself and the two tribes of Indians (one quiet, the other warlike) with whom he had to make his terms. This, he stated, with their full comprehension, to be a common sanction for right conduct in his and their firm belief in a Supreme Being who insisted on justice and generosity between all his human creatures. That this was his belief and not a convenient cover for gaining their confidence, he showed them by his conduct. This was the psychological agreement. He took care while meeting the Braves to begin first by stating his case openly and, while awaiting their reply, to conduct himself with that gravity and patience which were, to their excitable and repressed natures, the hall-marks of a responsible and trustworthy man. He used no subterfuge, giving them his full wishes and reasons and he attempted no haste, leaving them to deliberate whether they would accept his offer. The legal agreement was plain and just, and once there was philosophic agreement as to the ultimate sanction, and the psychological assurances that such a sanction really carried its full weight with the stranger, the terms presented no more difficulty than the seating of friendly guests at a common table. Once the treaty had been approved by the original land-owners, then Penn completed his introduction to them, and his establishment of himself, as a man who shared their values and could be approved by their standards, by showing that his physical training was such that he could hold his own in athletic contests with the younger Braves.

Here, then, we see in a classic and detailed report the features which go toward non-violent sanction and how the trained man is one who employs his intelligence to find a common philosophic basis, his intuition to find the behaviour and approach which reassures strangers as to his essential integrity, and his physical skill, the trained and athletic body, to prove and confirm in their minds the belief that the man they are dealing with so conducts himself not out of weakness or fear, but out of a common respect for them, for himself and the one spirit from which they are all sprung.

Penn's success seems, however, to have an alien early simplicity when we compare it with our tangled and inflamed relationships. We see a reasonable man of the oncoming Age of Reason meeting a number of Red Indians still little damaged by contacts with our low business morality. We need to remind ourselves that the Red Indians were fighting tribes in whom a strain of neurosis had made them to be reckoned among the most sadistic of savages and, on the other hand, that Penn himself, though he could and did with great constancy behave under the limitations of an English gentleman, did not draw his strength from such a strangulated and uncertain sanction. He had come wholly under the influence of the daimonic George Fox and his deeper consciousness had, at least in his early years, been put in contact with that profound and transforming psychological "field" which the early Quakers, of all the religious bodies of the seventeenth century, most surely and most strongly created.

Still it is doubtful whether Penn to-day would be able to contribute much to our present acute problem, or to extricate civilization from the quagmire into which it sinks to suffocation. We have to remember that his going to the New World was largely due to his seeing no way in which the standards and manner of life in which he believed could be carried out in the old. Men could perhaps gain private religious toleration but what use would that be if the State they had to serve stultified all their actions? If all their just dealing and wise and generous action only contributed to make a war-using and corrupt government the more able to use wrong means to attain wrong ends and to involve all its subjects in its disasters and its sins? As we have seen, the sweeping of a room in God's name may actually serve the devil's purpose, if it is but to leave the room "empty, swept and garnished" for seven worse ones. We have also to remember that though he did solve his immediate problem—to make the original people of the land freely acknowledge his position and possession among them—his psychological knowledge and training was not sufficient either to create a non-violent community among the Europeans he himself ruled, or to assure that this State of his should not later violate the very treaty provisions of justice and non-violence toward the Indians. The code of laws with which he supplied his State were in advance of contemporary Europe's, but they were pre-psychological, retaining capital punishment—which as Calvert established in his classical study *(Capital Punishment in the Twentieth Century)* has no deterrent effect—and many other penal regulations and having no recognition of that social prophylaxis, psychological training. Finally he permitted his son to succeed him to a position of such power that by chicanery with the Indians he was able to undo nearly all the work his father had done toward non-violence. To permit such a son to have such power was in its way, comparing the two men and their times, almost as great and criminal a negligence as Marcus Aurelius the Stoic permitting his debauched son Commodus to succeed him on the Roman Imperial Throne.

We see, nevertheless, how much can be done by a man trained up to the not excessive level which Penn attained. In the early Quaker groups his grave, courageous and naturally disciplined nature had found direct experience of *Charitism,* that passing "from life to death" through the intense awareness of the common love. Having discovered this common eternal love, and seen that it is latent in every man; however entombed in individualism it can be called like a Lazarus from the grave; Penn had enough sanction and sufficient power to make a society with a finer way of living, both in foreign and (though less) in home affairs, than any in Europe at that time.

But, as we have seen, when inquiring into the success of primitive Christianity and the failure of the prosperous sub-apostolic Church, Charitism is not enough, the spontaneous experience of small, simple groups if it is to be an equally valid experience of larger consciousness for the educated and the many, must be a definite way and process of knowing, a psychological technique. More than Quakerism is required to-day, but also a more advanced type than Penn (fine as he was) is now possible. We have seen that societies much more psychologically planned than Penn's Sylvania have existed—such as the Essenes—but their techniques of training are lost to us, and even if we had them we should probably need much effort before we could decode them into a form in which they could now be used.

The first step, then, would seem to be to discover and to apply all that the working animal psychologist, the trainer and educator of a dumb beast has now learnt. We are at this point taking for granted that we are wholly concerned with those who have, firstly, resolved their own internal conflict, and are no longer at war with themselves, finding that their own moods (even when others are absent or considerate) are incalculable and perverse; and, secondly, these are they who have learned to live in close and fruitful co-operation with like-minded people. The Doctor type, which springs from those who have already passed successfully through the two preliminary stages of training, is directing himself to master relationships and to take initiative in associations when those, with whom he is confronted, are either sullenly indifferent or actively hostile. Here is a situation in many ways closely akin to the animal trainer's, when he confronts a suspicious and dangerous wild beast. Argument is out of the question and were it possible, would only come too late. The trainer must gain the initiative. He must change the mood immediately, before any direct contact takes place. It is this attitude of mind which is beyond Penn's and which, if it is to be obtained and retained, needs careful and advanced training.

The first stages of this education are generally recognized now. The use of the surface will is perhaps worse than useless. Repressed fear and resentment signal, if anything more efficaciously than when they are openly expressed, to the subconscious mind of the suspicious or violent nature which we have to win. That is why people are often more at their ease with, and more fond of a passionate, uncontrolled nature than of one whose spontaneity has gone, whose cheerfulness jars as hollow and calculating, and whose control suggests not strength but how near hysteric breakdown lies below the surface and how fearful has to be its repression. Frightened of oneself, one can never reassure others.

The first step in training is, therefore, long before any such contacts would be attempted. The doctor-in-training learns to render his mind incapable of surprise and alarm, far less of suspicion and resentment. This is begun by removing from the mind all causes for such feelings. Such feelings are awoken by threats against the self's physical safety, its economic security and its reputation.

The doctor-in-training, therefore, beginning at the most abstract, newest and least physical level, aims at getting rid of all concern for his reputation. He aims at anonymity. No insult can thereafter affect him. He does not have to choke down any sense of slighted honour. He no longer has such a thing. It is as irrelevant to speak to him of his good name as it would be to speak to a frog about the tail that it once wore when a tadpole. The Neo-Brahmin must therefore have, as had the original Brahmin in Manu's code, no rank, position or place as an executive. He would be simply a number, an interchangeable unit in a field of force. It is, however, not enough to *say* we are free of an emotion. It must be completely atrophied in us, and practical tests must show that it has disappeared. The elimination of these spontaneous limitations is too serious a matter to be left to anyone's self-assurance, however sincere. We cannot take a man's word for it that he can remove tonsils, or pilot a vessel. He must be examined by experts before we dare trust our lives to him. Most people, who care wholly for doing so, can train out an undesirable emotion, but the only safe and lasting way of doing so, the only way which will not lead to repression, is through concentrating on and desiring the undesirable emotion's opposite. To get rid of an emotion without putting an equally potent one in its place is to find the will-to-live dangerously starved and to be in danger of the balked impulse, leading either to an intolerable sense of worthlessness and dryness—acedia—or to the energy striking down to lower levels, and violent, apparently groundless and certainly pointless attacks of sensualism or irritation being experienced. The doctor-in-training who is getting rid of any sense of reputation will do so by planning and concentrating so that all his interests are running toward a life which transcends individuality, a life in which individuality is seen as only a phase.

That is why only a philosophy which sees this life as a cross-section of some much vaster experience can really cure individualism. When we are rationally and emotionally convinced that individuality is only a husk and shell, confining and constricting our full life and experience, we will be able to avoid anything that would inflame the self, as spontaneously as the body reacts from being suffocated. The doctor-in-training who so lives will soon find that his self is increasingly becoming for him more a limitation than an expression. When such an one feels himself sufficiently free he can ask to be tried out in situations which might be expected to show whether in him there still lurk any traces of self-regard. It is very easy for anyone to discover whether the consciousness in anyone else is still strangulated or has already returned to its direct and unhindered flow in the vast current of Universal being. Until his associates are all convinced that the doctor-in-training has attained this freedom, he must remain in training among those others of his fellows who are also studying and practising to achieve this capacity.

The elimination of reputation and care for the self's recognition leads also to freedom from all other possessiveness. The wish for possession arises from a desire to protect and preserve the self. When the self is recognized with complete conviction to be a limitation, a frustration and not a fulfilment, then the wish for possessions atrophies. It becomes as absurd as the wish to be well known, to have fame. Two of the ties which frustrate the true development of consciousness are thus removed, and the man so released is proportionately powerful.

There remains the deepest, the purely animal, spontaneous, inborn impulse to save the body. Fear of physical pain, damage and death is the last and strongest of the roots which tie consciousness to its temporary selfhood. So it adheres to and becomes involved in the descent and foundering of what should be only the launching-cradle of the ship of the soul. Launched to float of its own buoyancy on an element which engulfs those stocks, the soul should rise above and leave them behind, as the tide of Timelessness comes in to carry it out to its proper life.

Reason can do much to argue and to laugh us out of our absurd sense of self-importance, by showing us that we are, as self-conscious individuals, always ridiculous and that fame cannot last. Beside a passion to be known and remembered, the passion for possessions is comparatively sane; for we can possess things for a time, but who has actually relished posthumous reputation? That many have thrown away actual if transitory comforts for this mirage is only one more proof of how obviously deluded and hypnotized are the men who call themselves realists. Reason can eat away our possessiveness with the same acid: "Soul Thou hast much goods laid up. . . Be merry. . . I will pull down my barns and build greater." "Thou fool, this night shall thy soul be required of thee." But against the spontaneous fear of pain and death, a deeper treatment is required. Not a different one, however. It is the same treatment carried down to those deeper levels, where these subconscious cramps and ties are locked. We now know that common standards are no more and no less than group-suggestion, mass hypnosis. We now realize that all our life as human beings is held together not by clearly calculated, mutual self-interest but by these suggestions. We now know from anthropological study that there is practically no suggestion, no pattern of living, no custom, rule or tabu that people will not accept and cannot bring themselves to act upon cheerfully, even to the cost of acute physical pain, privation, mutilation and death. We now know that to obey that suggestion, to have the sense that that command has been fulfilled is sweeter than life itself, than any physical reward. It is not pain or privation that drives a man to suicide (as Adam Smith observed with wonder a century and a half ago) but the loss of his fellows' approval, the sense that he has failed to fulfil the order which his subconscious, suggestible mind had accepted. This fact is now generally recognized and generally awakes despair.

It need not do so. Why it does so is because it was first rationally recognized (it has been subconsciously known to every leader since the rise of consciousness) by rationalists who thereupon assumed that all intentional conduct would henceforth become impossible. Some indeed went further, maintaining that, since this discovery, it was clear that all intentional conduct had always been an illusion and its appearance was due only to rationalization, to a pretence of reason disguising from the self the essentially compulsive and irrational nature of its actions. But the mere fact that man is so widely, sometimes it would seem infinitely, suggestible; the mere extravagance and absurdity of some of the inhibitions he has observed and actions he has carried out, prove beyond doubt that he is pre-eminently malleable. The subconscious mind, even at the lowest rating, will certainly accept anything which is presented to it in the form and through the method it comprehends. It will accept a reasonable proposition as easily as an unreasonable one, with only this difference from the assent given by our reason, that this deeper mind, having accepted a proposition, acts upon it and acts with complete assurance, energy and selflessness.

This is all the commonplace of ordinary psychology, the agreed basis of every successful school of therapy or analysis, however much they may diverge and differ after this point of common departure. And it is a firm and most adequate foundation on which to build this new education of a new doctorate. The process whereby fear and pain may be removed is by autosuggestion. The process is and must be a gradual one. Nor will it be an abiding one unless it is known to be true. That is an important proviso. All suggestions, if given correctly, will in time be accepted and acted upon with what appear to be complete conviction and ultra-rational power, but only those will last indefinitely which can be accepted as true—that is, when the consensus of experience—what one is told that an experience means and what that experience turns out to mean for oneself—agree.[1]

We are, then, only at the beginning of what we may do. Suggestion is but a preliminary shadow of what auto-suggestion can and will achieve (and has already done in singular cases) when it presses upon the line of cleavage through which the confined psyche must emerge into its fuller life and further evolution. But already we know that such auto-suggestion, directed positively so that the person lives continually in his supra-individual life, can dismiss fear and inhibit pain. The new intense quality of attention, now at the trained mind's disposal, draws away all the raw energy which could discharge in pain. Pain is, in fact, only a symptom of balked evolution, and as the whole idea of this new training is to continue evolution, the very source of pain will be tapped and pain itself must disappear.

The doctor-in-training will, then, achieve by a way of life and by constant mental exercises, complete freedom from the self. He will have the complete anthropological approach. Already it has been established among the savages, in the wild animal's den, in the luna-

1. This question of the quality of the conviction which the hypnotized person feels, has long interested researchers in hypnosis. It is commonly known that a hypnotized person told that a turnip is a peach will bite into the turnip with gusto; X-ray photographs taken of a hypnotized person's stomach when completely empty showed the stomach contracting but as soon as the suggestion, "You have had a full meal," was given, the contraction ceased and the stomach relaxed as though it had been lined. Yet hypnotized turnip-eaters, questioned as to what they were eating have been known to reply "Turnip." Milne Bramwell, the first London specialist to revive respectable hypnosis, carried this exploration in another direction. It had long been held that a hypnotized subject would never accept any suggestion which outraged the accepted sense of decency or morality. Bramwell told a subject under hypnosis, as a post-hypnotic suggestion, that some sugar was arsenic, that a colleague was coming to tea and that at tea the subject would put the white powder in the colleague's tea. The subject (having of course awoken from the trance and having no conscious memory of the suggestion) came in to tea and insisted on giving the colleague a helping of the white powder. The next day under hypnosis the subject was questioned by Bramwell as to why she gave the colleague arsenic and replied immediately: "It was not arsenic but sugar." The subconscious would seem, therefore, to be able to understand and to discriminate when a suggestion is given. The thing it must do is to carry out the spirit of the command; pretend to like a turnip as though it were a peach; hold open the stomach as though it had a meal; go through the pretence of giving arsenic. In each case it looks as though a better result could be achieved by telling the subconscious, without any pretence, what is required, and leaving it to make the psychophysical modifications. Then what must otherwise be only a temporary accommodation, a piece of immensely convincing play-acting, will become first an accepted pattern and then an organic adaptation.

tic's ward and the criminal's cell, that even an unconscious and unintentional degree of interest, fearlessness and integrity will win the suspicious creature's confidence. Such discoveries were made almost by accident because the old alternatives of violence or argument, compulsion or reason were for the moment out of the question and something had to be done at once. They are still only emerging from knacks and notions into the level of a clearly understood and plainly taught technique.[2] The Neo-Brahmin, coming as a culmination of a new education which reunites the consciousness, and rebuilds the community, the co-operative group of unlimited liability, will be, with his doctorate powers, able to bring together, as a single intentionally achieved and universally applicable science and technique, all these rudimentary knacks and methods.

To postulate that, is to do no more than to assemble and combine a number of techniques all now worked out as ways of handling specific acute problems in human conduct. The Neo-Brahmin, on this showing, would be no more than one who in himself sums up and knows consciously what the alienist, the criminologist, the animal psychologist and the anthropologist all have to know, though few of them could express it. He makes from the scattered fragmentary knowledge a system which all can learn who will train, and which can take the initiative in every human problem: individual, social and international.

With this alone, with nothing but this degree of power and appeal, we cannot say that a new force would not be entering into the world, a force adequate to answer the desperate appeal for a new, non-violent, non-self-defeating sanction. Scattered and unconscious, this force can only win local victories, to have all its detailed successes stultified by the general political degeneration, as to-day the best economic and productive methods, the best hygiene and housing, only make the competing nations more powerful, only feed the all-consuming fire of war. Conscious, comprehensive and united, here could be the New Force, the apt and appropriate energy which could

---

2. See Dr. White's *Autobiography of a Purpose*, Doubleday, Doran & Company, for a description of this research in the therapy of the insane.

salvage civilization, order humanity, give each degree of consciousness an adequate station and an adequate productive and expanding aim and show to the leaders the meaning of existence. This summation of all these scattered emergent techniques would itself be (as is a chord more than the sum of its constituent notes) something more than a psychiatry, more than a psychology, more than an anthropology. For it would have as its aim something more than the wish to understand the creatures it was studying; it would know not merely what they actually are but what they can become: it would understand that these creatures are an essential part of the being which must embrace and combine Knower and Known. This knowledge would also prevent those of the new outlook from having as their aim any such restricted and stultifying objective as returning the deranged and the dislocated to "normalcy." Those who are cured must be put under those new conditions, as well as given the new vision, whereby they may develop to the full their present potentiality. No aspiration of which they may now be capable, must be thwarted or blunted.

Such a comprehensive and dynamic attitude toward contemporary life needs a name which may combine two initiatives, that of service (compassion) and that of science (understanding). We may, therefore, name it provisionally the Philanthropological approach.

At this point there would seem nothing more to do than to suggest how such a doctorate Force might serve the world, what steps it would take to demonstrate its capacity to solve civilization's problem through an international philanthropological civil service, how it would set about to contact and control our present convulsive international anarchy.

Honesty, however, compels a further description of the Neo-Brahmin's powers. At present there has been described no more than such achievements and equipments as any successful alienist, anthropologist or animal educator has to attain. That such powers, adequately understood, ordered and directed might be sufficient to set our chartless and capsizing civilization once more on an even keel and on its course, is not a sufficient reason for not stating the full capacities to which the highest type may attain. We may need far more force than we estimate to prevent civilization foundering. The most advanced anthropological initiative may prove helpless against the modern mass neurosis, as rationalism has been helpless. No one could have foreseen such vast fanaticism, so ruthless and destructive; possessing peoples so informed and powerful. The tide has now run for years in this direction toward the brink of utter chaos, and no one can say that it shows any sign of slackening but rather, it must be owned, it is gathering volume with ever-increasing speed. We cannot, therefore, afford to underrate or to neglect any powers of which we might avail ourselves. We must also remind ourselves that the power to swing civilization on to another course will depend greatly upon the impression which the highest type of man does himself, by his character and capacities, his being and power, make on ordinary men. The success of the destructive enthusiasms of today, of the crusading strangulated groups (whether they be horizontal or vertical), is largely due to the fact that utilitarian democracy has not been enough. The human psyche, especially when it is young and vigorous, desires to have some great and difficult objective. Permanent prosperity is simply the promise of permanent boredom, for simply to increase means in default of adequate ends is to choke the small sense and spark of purpose that man naturally had and which the struggle for adequate means, did fan. The Romantic Movement which tried to satisfy the young (before they grow old and care only for safety) with the notion that being in love with an individual and securing his presence is sufficient to make life "happy ever after," is so feeble as only to serve to illustrate utilitarianism's psychological bankruptcy.

We have seen that this spectre of futility and frustration—the boredom that drives men to drink, drugs and the mass murder of war—always haunts mankind—a Tantalus horror snatching comfort and satiety from him when he thinks he has them in his hand. This must be so. We must repeat: the State Idolatry, the worship of a cause in which the individual is consumed, must come, always forestalling that material prosperity and utilitarian peace which we pursue. For that prosperity and peace is a mirage. We are not made to be permanent individuals. Our evolutionary purpose and task is now to emerge from, and transcend our individuality. Hence, rather than accept the offer of anyone who would gild, however heavily, our chains, we, like the Gadarene maniac, rush headlong to dash off our bonds, even at the price of breaking ourselves and destroying our neighbour. Therefore in this attempt to put mankind again on its evolutionary course, nothing is more important than to give those of mankind who dread above all being permanently imprisoned in the self—those people who must be the prey and power of the Fanatical Movements mastering the world to-day—the vision of the true emergence, the example and image of what the New Man can be. Against the modern false prophet must be raised up the true.

And the false prophet has done one thing for us. He has made us able to know at least what we may and must expect of the true modern prophet. The powers of the new leader of the present fanaticisms are at least as interesting and as unexpected as they are dangerous. Here is a psychological energy and quality which would have been as unbelievable to the rationalists and the Utilitarians as X-rays or the Devil. Here, if only in an evil and destructive form, is a psychological power approaching to some equality with our physical powers. This, as has been said, is the daimonic person, a man whose tremendous influence does not depend on intelligence or rational understanding, or self-control or self-knowledge, but on the possession of a gift (or the subjection to an influence) which yields him mastery over the violent, the cunning, the capable, the gifted, the healthy, the generous, the devoted—in short, over all who come under his spell and whom he wishes to win, save the man who has understood the meaning of life, acted upon his knowledge and so can know to the full the fanatic's power and yet wholly discount and dismiss it.

The time then demands nothing less than the positive creative opposite and equal of the destructive and daimonic leader. Such the philanthropologist is not. He is much. He can solve many difficulties and resolve many conflicts. He can turn ordinary ill will, suspicion, resentment and violence into creative channels. He can do far more than can any utilitarian, depending on demonstrating how poorly conflict pays. He can bring about agreement between conflicting social groups, each of which has much right on its side and considerable force at its disposal. Against the daimonic type, however, even he becomes ineffective. Where there is some basis of a common social solidarity, the anthropological approach can avail. It can gain purchase where the mind of the stranger has only native suspicion blended with native curiosity, a sense of misgiving with a sense of his own rightness. When the natural misgiving has been deliberately distorted and inflamed by a dreadful prostitution of the anthropological-psychological methods, by the super mass suggestive power of modern propaganda, then the philanthropologist is ineffective, his approach fails. This kind goes not out save to a power equal to itself. Yet beyond the philanthropological equipment there are endowments to be had, equal to and surpassing those of the leader of the nationalistic neurosis.

We need not despair that there may arise the adequate reply to the daimonic leader. He appeals not merely to what is atavistic in modem man but, as we have seen, to passions which are neutral and might be progressive. Man is a creature which has not merely suggestibility but a past (still existing in his subconscious) in which powerful emotional appeal made to his group "instinct" was well-nigh irresistible. But man has also a future as well as a past, and the skill of the Dictators is that, with a single appeal they catch both their subjects' atavistic and decadent longing for security, exclusiveness and dominance, and also their subjects' rudimentary but progressive longing for a great cause and future in which they may lose themselves, embrace mankind and serve the general meaning and purpose of life. The smaller but far more powerful side of their subjects' mind could be split away from them, for it only belongs to them by default. That arrested individualism which expresses itself in materialistic Liberalism will never be able to rally to its cause the creative, progressive force which the Dictators are successfully exploiting. A new prophetic type would do so, and must.

It will be said, no doubt, that this is a vain hope. The gigantic appeal which can be made to the masses of mankind, whether it is to persuade them to attack Jews or smoke tobacco, can be made because we wish profoundly to indulge in dislike and in smoking.

The spellbinders play on our latent uneradicated weaknesses. They are our baser nature finding an external commanding voice. They could never make us accept with enthusiasm a course more lofty, instead of one more base, than our norm. This, however, is only an assumption. Liberal democracy certainly has not used the appeal and the technique of the modem spellbinder. Beside the skill in popular appeal of a Dictatorial Director of Propaganda, how rudimentary and feeble are the arts of the most effective demagogue. Emotional appeal to the subconscious, to make men live a higher life, has been used by religious revivalists in the past, and no one who reads in Benjamin Franklin's journal how that staunch rationalist felt his resolved power to inhibit his almsgiving melt under Whitfield's spell, can doubt such powers are neutral and could be used well. True, religious revivals are local (though some have covered considerable areas and perhaps the modern use of the loudspeaker would permit them also to sweep whole countrysides). They have also proved temporary, but then that, too, will quite probably prove to be the fate of the Dictatorships. Suggestion always has limits and after a while, unless steps are taken to confirm, build up, edify those who have been melted, by giving them a cast in which to form themselves, they will relapse into their original irrelevancies and incoherencies. As we have seen, a hypnotic suggestion can only endure if it is afterwards confirmed by other apprehensions and so leads to functions which in turn create organic alterations.

If, then, we so wish it, the new leaders who will lead to the future not to the past, to peace not war, to life not death, can arise. For this Neo-Brahmin, the new prophetic type and forerunner of the succeeding world order, we see that we can postulate already three things of immense force: (1) a height of integrity beyond that which any save sporadic individuals of the most advanced type have hitherto shown; (2) a clear understanding of the meaning of life, of the direction and the rate of the present crisis of evolution; and (3) a power of appeal which has been demonstrated in its destructive aspect by the modern Dictators.

These three powers can find purchase in the modern spirit, for we know that it was from need of such an appeal and through lack of such inspiration that the best of the young turned to the false prophets, the Dictators. Nor will such leaders be lonely individuals, solitary and hardly knowing how they came by their vision and power; they will not only be a numerous rank supporting and controlling each other, but they will be products of a definite, scientific, clearly understood method. They will be able to be made as armies are made, by calling for recruits and training them; as queen bees are made out of ordinary worker grubs when the hive has need, by giving the common grub a special diet.

But again, we cannot in honesty stop here. These endowments, considerable as they are, and, it may be, adequate to salvage civilization and put mankind again on the path of its true evolution, are not the full gifts which such a type as the Neo-Brahmin may possess—which, indeed, he must manifest if it is true that evolution can continue, and will continue as psychological development. We are not returning and we cannot return mankind to normalcy or to collective security for the *status quo*. The hypothesis of this essay is that evolution is seeking its only possible continuation along this path of enlarged consciousness, and that those who have grasped this fact can not only co-operate with it, but that only by their co-operation will further evolution be possible. Further, the evidence collected in these chapters seems to indicate that by understanding the meaning of human pain and sex we come to the realization that in our pain and sex we have an indication of the sum of still unused evolutionary energy yet residing in our psycho-physique—the accidental escape and discharge of which gives us the slaking and even stupifying return to contentment, which is the aftermath of lust. Finally, the evidence seems to have indicated that sporadic outcrops of new extra-sensory faculties, a purely psychic evolution, are not the only indications of the direction which our further development should take. Psychic evolution can be, and will only be, developed out of that remaining store and sum of psycho-physical energy which, if it cannot be given its evolutionary release, we either squander in eroticism (and indeed in all sexual discharge) or are distracted by its spontaneous leakage, in all the degrees of balked distress, from irritability, anxiety and hyperæsthesia to the horrors of deliria and the agonies of spastic pain.

The Neo-Brahmin, by developing these sporadic powers, through mental exercises and by feeding them out of the reserved stores of basic energy (by continence and diet) brings such powers under his conscious control, so that instead of their functioning sporadically, they can be produced at will. When we consider the enormous influence which has always been exercised by men who had no power but their absolute integrity—how Herod feared the Baptist, how Ivan the Terrible dreaded the fakir-monk who day by day denounced him in the Red Square of Moscow, how Cromwell was disturbed by Fox, how the Caliph listened to Francis—there can be no historical doubt that the Dictator type, the daimonic man, is if anything more vulnerable to the charges of the fearless mystic than is the average sensual man. For the daimonic type and the militarist, always concerned with real force, if only with the fringe of it, know that their power and success are but ripples on the shallows of a deep out of which at any moment may come a tidal wave washing them away and obliterating every trace of their achievements. Add, then, to this integrity a power to demonstrate in contemporary language what Life means and intends, to show the fate of every man which he cannot escape unless he fulfil it, and, then again, to this power of delivery and this clearness of message add the extra-sensory powers, the power to know the thoughts hidden by the crafty man in his mind, the power to know what men are planning and contriving elsewhere, and here is a Force, a unity of enlightened men, which at last, and such only, would be adequate to forward evolution.

Such, then, is the final rank of the New Order, the contemporary college, the neo-monastic answer to our own Dark Age. This is the type of doctor which would be the full term and development of the new dynamic caste, the Neo-Brahmin rising out of the Neo-Kshatriyas, the mastership administrator rank, as that again springs from and leads the Neo-Vaisyas, the novice technician rank. This Neo-Doctor, having no property and no executive power, would be the incarnate conscience of mankind, the inspired actual sanction of conduct which can, and alone can, lead humanity out of its impasse on to the further slopes of evolutionary progress.

That such a type is a possible development, that it is evolutionarily valid, that we have the power within us to make this forward step at Life's call and challenge, that this is the moment in Life's growth, in humanity's advance, in civilization's explication when such a mutation must be made, must be made in consciousness and if not made, then our precarious, over-reaching effort (failing to find its keystone) must crash, there seems no further reasonable doubt. Where else, save in the most pathetic wishful thinking, lies any gleam of hope on our own horizon? Is it not daily more clear that the whole structure of civilization sways to its fall as we add with feverish pace still further disbalancing physical and menial forces against an ever more inadequate and shrinking psychological knowledge and power?

Nor is the ordinary man unaware of this state of affairs; nor does he neglect, like Bluebeard's wife, having won a few moments' reprieve, to call out to any watcher, "Is there any sign of help coming before I must go down to my butchering?" The world is aware that physical violence and the pretended sanction it claims to give, are about to destroy civilized life. The average citizen knows that armament not only sucks dry his social resources, but is an instrument of increasing inaptness and imprecision. Applied physical science, in all its other manifestations, has given men instruments of precision. Whether for better or for worse, at least men got precisely what they demanded, as never before. But in this respect warfare alone has been a sinister exception. In an ever more complex and interdependent civilization it grows increasingly important and vital that force when used, and, even more, violence, should be precise, limited, defined. Instead of this, when our civilization is most involved and interrelated, war has become increasingly pervasive and irrestrictible—as little to be controlled as a fire started in a haybarn by a hobo trying to smoke out a fellow tramp and trespasser.

This problem of appropriate force has then acutely concerned practical men, men none of whom imagined either the present precipitate descent of civilization or the possible hopes of further evolutionary development. Nor did this interest in the task of finding a method of keeping agreed order begin with the twentieth century, and the mechanization, and the rapidly increased dehumanization, of war. The rise and development of the modern police force dates from a century ago when in England in the thirties, the Reform Bill had enfranchized many till then considered outside the bounds of the governing classes, and the tolerance and decent dealing they extended to their own kind. These new constituents of the government could no longer be considered as the "mob," rabble, who, when they rose against their masters and their miseries, could be shot down by the military. Yet they could not be considered as equals. There was between them and the ruling caste no common unconscious tie made, by common lineage and education, to create a spontaneous understanding and an unquestioned, unabused tolerance. Two classes were meeting, classes which observed different conventions, sprung from different traditions. The use of unrestrained force was out of the question; the invoking of civil war, the employment of the utmost cunning and violence, had been ruled out when the Duke of Wellington with his immense military prestige announced from the premiership the capitulation of the aristocracy in the words, "The Government must be carried on." Civil war had been renounced by the minority, what we should call the Fascist class; social solidarity had won. But such solidarity was far less of a fact than a promise. Some kind of "Force," half-way between the unquestioned force of an almost unconscious tradition and martial law, had to be created. The police force was Sir Robert Peel's answer.

This was no sudden invention but, like most devices, an evolution. A constabulary had, Peel discovered, to be devised to hold Ireland. Eleven years after, he modified this force to fit the task of controlling not an alien agricultural people but that urban morass, London, rapidly becoming the largest, most planless city in the world. The police force, from its inception, was a developing, provisional body devised and modified to suit varied and successive problems of social order. It is the shifting compromise, the make-weight and rider that is moved from the one end of the scale of Force to the other. It runs all the way of the moral spectrum, from cruelty which would disgust the soldier to patience he would despise. At the lower end it is of the utmost violence and cunning, the secret police, which by use of the spy and the informer and by the invention and employment of every refinement of physical and psychological pressure, distress and torture, corrupt the average citizen, paralyse the good, and give government into the hand of those who are least capable of pity. At the upper end it is the anthropological organization which exists to reduce friction between individuals all almost wholly concerned to attain their private ends and to permit expression between social groups whose ambition to master the State makes them disregard social solidarity and the equal rights of other parties. During the nineteenth century the evolution, the necessary evolution, of the police force was not thought out, probably because there was a common assumption that order and the growth of social solidarity would, through increase of wealth and intellectual education, gradually effect such social progress that even a police force would become otiose. Now we have learnt, what the study of evolution should have shown us, that development can be down as well as up and that the older the organism, the more liable it is to become decadent than to progress, that indeed, at the point we have reached, our organization can only escape degeneracy if it can willingly adopt intentional progress.

The modern Dictators, we need always to remind ourselves, are ruling, and are unprecedented in and by that rule, by police forces, highly advanced; terribly, effectively decadent. These new despots are the first men to master great states without being "masters of the Legions," without returning, as did Cæsar, Cromwell and Napoleon, victorious generals at the head of their loyal armies. The new tyranny is newer than the most modern democracy. It wins power by ruling waves which the young Jew, Hertz, was first to detect, hardly a generation ago. Through the mastery of these radio waves it beats down the will of its subjects and like a tide percolates, spreads, surges over, and sweeps away the concrete and steel defences of its neighbours. Here is direct action indeed. Here is that concentrated and focused attack on the will of the enemy, which Clausewitz taught should be the sole objective of a military commander. Here is precise and really contemporary cunning and violence. And once the victory is won, then the occupied territory is held not by soldiers but by the secret police.

Yet, while the police force itself is, throughout the world, undergoing a degenerative development, because it has no philosophy, no teleology and can only conceive of itself as the modern and apt development of militarism, the scientific development of violence and cunning, at the same time the administrative aspect and growth of applied anthropology is extending. The organized police in India, and even in the United States, Britain and Switzerland, may be inevitably developing toward being Secret Police, using the informer and the third degree; becoming the master and not the helper of the citizen. Nevertheless, as we have seen, the anthropologist, the criminologist and the psychiatrist are still advancing our knowledge and our application of, our science and our training in the new technique of Power in the New Force. When to these three branches of research and therapy we add that of the animal psychologist, the man who reconditions, or rather deconditions the beast, and who, without argument or logic, educates the dumb brute out of its brute prejudices into intuitive understanding and co-operation, we cannot fail to see that the conception of a New Force, a new training, a new inspiration and a new sanction is ready to dawn on us.

Briefly then, the world is looking for an International Police Force; and in this conception and the actualization of it (which humanity would gladly give, if only there could be found men adequate to fill its ranks), the Neo-Brahmins will find their channel, their cadre. But that phrase—International Police Force—we must rid of serious misconceptions forced on it by Yuen who, through long association with the decadent forms of international relations, could only think in terms of more extensive militarism. The term means for most people a League Air Army which the possessing nations—a tribunal shown to be incapable of impartiality, let alone anthropological insight—should empower to enforce its all too human decisions by that least precise, apt and just instrument, the bombing plane. A truly contemporary and apt World Police Force would require three things to make it effective. First, every member of the Force must be loyal to humanity, and to nothing else. This elimination of all partial and strangulated loyalties, to one's nation, state, territory or class, cannot be achieved by an oath, an argument or a uniform. It can only be effected by a training such as we have seen the New Doctorate would undergo. The old Leagues and Imperial Governments, as we know, could never trust their executioners to execute their sentence against fellow countrymen. The Irish Constabulary had always to be drafted to a different district from that in which they grew up. Secondly, each member must be free of every personal tie and physical weakness. Thirdly, each must have achieved to the full the philanthropological approach. Once there is such a body of New Doctors or Neo-Brahmins, men combining the selfless zeal of the best missionary, the tireless interest of the finest naturalist, the fearless patience and affection of the most effective animal educator, the aversionless understanding and insight of the most successful psychiatrist, here and here only is the New Force, the psychological power adequate to balance our prodigious physical energy. Add to this equipment the fact that these men will be truly prophetic, men of real vision, who see into the minds of the daimonic leaders as clearly as into those of the led masses, who speak to the condition of both sorts because they exercise the absolute conviction of the seer, whose understanding insight and realization of life's meaning can be as little in doubt as his integrity, who are known unmistakably to have powers which flow from such knowledge and which confirm that knowledge for the common man by giving him clear sanction for right and evolutionary conduct—then there is, indeed, a hope for something beyond the salvaging of civilization. There is, under such leadership, the expectation that mankind as a whole could resume its advance. Instead of a remnant being saved, a few pioneers getting through to the new level, the whole vast column of mankind from the seer, through those of the avowed life, past those who at least wish to heal themselves and be individually whole, down to the many who can at present only wish to be happy and would prefer to be kind, the whole vast spectrum of incarnate minds, the complete range of the psychological types of mankind, might be in its entirety shifted upwards toward an increasing awareness, an increasing apprehension of their common unity and of the comprehensive purpose of their existence.

Such a civilization, such a united humanity would have in every country and district these new colleges. Into them would be entering continually those who came in order to understand their acute and challenging need for self-integration, the full education of the psyche. In them would be living those who help the seekers at that first level of education, who were educating themselves still further and were giving rise, from the level of Mastership, to applicants for the Doctorate. Out of these colleges would be continually passing those who had become thus proficient, passing out to undertake in every way the new Civil Service to mankind. Bands of such—as the white cells flow to any part of the body threatened by breakdown—as large, free-moving, protective cells are always found at any possible focus of infection—would move at once to any place where tension was rising. They would diagnose the trouble, explain the friction, allay the suspicion and resentment and indicate the solution. As did the early Friends, under a spontaneous impulse, these perfectly emancipated men of a New Friendship would pass (free-moving vital cells) carrying inspiration, insight and understanding to those points where through some infection or strangulation, vision had become lost and mortification was threatened.

Such then, is a possible hope for mankind. It is a radical departure from any ordinary expectation. It would be in fact an evolutionary mutation taking place in our psycho-social evolution. Yet we know that evolution has advanced by such decisive steps, by Life arriving at a crisis out of which only a profound creative act can extricate it, and where, unless such an act is achieved, the future can only be cataclysmic collapse or miserable decadence. It is unmistakably evident that humanity has arrived at such a crisis, that that crisis is psychological, and that there is a number of clear if not coordinated indications of the direction in which a new advance could be made.

Will it be made? We cannot say. "You can take a horse to water but you cannot make him drink." You can throw open the doors of a burning stable but you cannot make a horse come out into safety. You must go in, soothe him, melt the paralysis of fear which is holding him rooted where he will be destroyed, and if your integrity is sufficient he will follow you out from death. We cannot say, even if we who care can make ourselves sufficiently trustworthy, whether humanity will follow. No violence, not even the most indirect and painless of coercions can be used. The only possible meaning of life is that here, under Time, human consciousness discovers itself. The Universe exists for the emergence and development of free creative consciousness. That being the meaning of Time and the phenomenal world, to coerce individuals to make them behave and to compel them to do right, is to frustrate their growth and make futile the evolutionary purpose in that particular. If men can think that the purpose of life, or the only attainable purpose in life is a Utopia where, after an unknown number of generations of fine men have suffered and committed crimes to achieve it, there will be a conflictless, painless, appetitive world filled with aimless populations scampering about, feeding, sleeping, courting, procreating and dying, without a notion what the whole thing means and why they are born and why they die; huge droves of animals handicapped in the enjoyment of senseless pleasure by an awkward relic of intelligence, curiosity and wonder, waiting, turned out to grass, until the cosmic frost cuts them off—if anyone can believe in such a self-stultifying daydream no doubt he may use, as Communists believe, any means. It does not matter, if you stultify man's power of curiosity and wonder, his need for liberty, his right to experience, if his consciousness is insignificant, if his mind is merely an epiphenomenal accident, excessive steam given off by a physical engine when it is kept from running about. But if consciousness, ever more awareness and the increasing apprehension of a supra-physical world, is man's only adequate aim and goal, and is the aim and goal of Life through him, then any attempt to drive him, even toward awareness, must defeat the end.

We cannot, then, say that if we provide the service, mankind will choose to avail itself of that service. The Buddhist saying, "All the Buddhas point the way but because it is the way, everyone must tread every step of it himself," has in it that realism and patience which marks the highest oriental thought and which is so hard for the febrile occidental to realize. Indeed, we must go farther. However hardly we strive to see the future, free of our present material prejudices, and to understand how minds more advanced than ours would think and act, we cannot be at all sure as to how the world would appear to them or what line of conduct they would pursue.

"If I were God" means only if I were omnipotent. But were I omniscient also, then my utter power would be matched and balanced by my complete vision. I should see things *sub specie aeternitatis,* as Spinoza knew we must all desire, if we would see them as they are, but which we may never attain to till we are, at the very least, passionless. The question as to whether the new higher type (into which men must mutate if they would carry on evolution) will really alter the main stream of history and give humanity a future, can, perhaps, best be answered by another question, crude but crucial: "If we were God would we do more than God does?"

We can make preparation. We can get ourselves ready by transcending the self so as to see things as they really are. We may hope that in the end all frustration and futility will be resolved. We must not, however, rule exactly where and when and how that can be. To attempt to do more than to emerge and be prepared is to be like a chick in the egg, when the time has come to break its way out, losing time and energy, which are for it both strictly limited, to pause and speculate about the vast world outside and how to behave once there. So the simile of the egg again returns and closes us in. We must turn to our work, our actual effort to work our way out, to attain Liberation.

CHAPTER XVI

SUMMARY AND CONCLUSION

IN THE foregoing chapters the following thesis has been developed. There seems clear proof now that Evolution is culminating to-day. It may seem ridiculously provincial to suppose that the gigantic secular process of Life's emergence can move appreciably (still less suddenly fulfil itself) in our moment of existence; "as little might we hope to detect movements in the fixed stars." Yet this fear of chronological provincialism is itself provincial. Truly detached observation shows that the tempo of growth has so accelerated and can so accelerate. It is only a nineteenth-century prejudice against rapid change and a wishful belief in a fabian regularity and gradualness that disregarded the cataclysmic and mutational element as the decisive force in all change and growth. The "curve" as well as the pace of evolution confirms the belief that we confront and are part of an evolutionary crisis. We now see that we have to correct here another prejudice of men who thought they were detached from and could observe from without the world of life. Our history is part of natural history. Human history is biology, psychologically accelerated. The transitional term between the natural history of *homo* and the civilic history of humanity is the Old Stone Age. It matches, in its time span, its tempo and the quality of its achievements, as the middle term linking up zoological evolution with human discovery and invention.

Specifically human history has been from craft skill to speculation, from an *ad hoc* apprehension to general comprehension, from applied knowledge to pure mentation, from understanding in order to control, to understanding why the self wishes to control, from power over environment to contemplation of the seen and the seer as one. This is a rapidly accelerating inturning process which may be plotted under three heads or points: history may be seen as turning from the development of the physique, through the development of technique to the development of the psyche.

Such findings make it possible for us to accept only one conclusion. We are at the End of the Age, through which evolution has extended. Faced with this conclusion many may decide that man is decadent; there is no further evolution possible for him, and he will either fall into degeneracy or rapidly destroy himself—the latter seeming the more probable.

Yet man could be saved, as Life has saved itself before, by a sudden and radical mutation. True, he is a gigantic ape evolved from much smaller ancestry and "no giant gives rise to a giant." True he has made, at last, a definite specialization, a huge hypertrophy, and that, nearly always, is a sign of evolution's termination in such a stock and after such a specific change. But he might continue his evolution mentally, not physically. In support of this we know that man has not modified structurally for a million years or so, and that is a considerable period of arrest for a member of the late mammalian species—the elephantine mastodons went through their entire specific evolution in four hundred and fifty thousand or five hundred thousand years.[1] We also know that in that time and with constantly accelerated skill and advanced mental power, man has developed his powers by technical progress. Suspension or arrest of physical evolution has been accompanied by a unique extension of mental evolution. Similarly, the definite specialization, the huge hypertrophy, is precisely the brain, which not only is the organ through which mental evolution can express itself but is the largest mass of undifferentiated tissue in the body and which undoubtedly functions not as a series of wholly specified parts, but as a whole.

It would appear, then, that man has, physically speaking, specialized in unspecialization and by this means has won himself a new span of evolutionary life and development. In this new span he, suspending physical change, carries on further development, but also in another medium. He advances mentally or technically, by modifying his environment to suit his needs instead of, as heretofore, altering

1. See "Pleistocene Chronology," Dr. Freidrich Zeuner, *The Geological Magazine* (London), 72.19.350-376, 1935.

his physique and its needs so as to fit in with his environment. So he wins new extension of evolution, but that change has henceforward to be with increasing speed and increasingly psychological. Man is now not merely a creature balancing on a rolling ball; he is balanced on a rolling ball which as it rolls leaves the ground and, like a meteorite, plunges through the sky. On the earth you can stand still. If you take to the air you must move if you are not to fall, and the higher you rise, because of the air's rarity, the faster you must travel.

But granted that to-day something even more radical is demanded of him than even enlarging powers over his environment, that in point of fact, these powers are now threatening him, as the hypertrophy of horns and tusks have threatened and destroyed the animals which grew these weapons and tools too freely, can man to-day make as radical a change (this time a change in consciousness) as radical at least as that which he made when he rose from animal to man? In support of that hope we know now that whereas every other animal species seems arrested for millions of years and to have entered on a decline, man alone seems capable of change, and, for the carrying out of that change, he alone seems to have an immense store of still unused, undifferentiated primal energy, for we know that his hormonic charge is higher than any animal's and we know that his sum of sensitiveness and his nervous tension is also the highest of all beasts. We know this through the fact of his capacity for pain and his enormous sexual capacity—a capacity in itself, as sexual energy *per se,* quite unnecessary to a slow-breeding, carefully-rearing animal.

This store of energy, we also find, can have appropriate advanced mental channels of expression; when so used, lust ceases to trouble man and, when higher and more specifically psychological channels of expression are found, man ceases, not only to be sexual, but to be capable of pain. The highly developed intellectual type tends to find when in complete intellectual absorption that he. becomes indifferent to sex, and the practice of contemplation makes that freedom perpetual. The development of further psychological powers, such as extra-sensory perception, leaves the individual free of all acute physical sensation.

For some four thousand, five hundred years man kept his mental advance roughly balanced (while his bodily development remained stabilized). During that time his progress in knowledge of the outer world was gradual but it was never completely arrested by collapse, for it was generally balanced by his advance in knowledge of the inner world: physics and psychology equated: means and ends were proportionate.[2]

Then (perhaps owing to a slight disbalance provoked by some disregard and increasing neglect of the necessary stabilization of the body), psychic growth, psychological advance became arrested and after three or four generations, advance in physics, through analysis and finally through mechanism, swallowed up nearly all of man's conscious creative energy. Then man's psychological condition became acute. For, not only the mind which had knowledge of the psychic aspect of reality became forced down as a subconsciousness and sealed over with an impermeable limen, through which the conscious mind could hardly at all reach it, but the foreconscious became a strangulation of consciousness and thus lost profound touch with reality and direct intuition of and with its fellow minds. Hence, an immense increase in force and complete paralysis of grasp, an indefinite extension of means and a disappearance of values and ends. Hence, we have to-day a society which must capsize unless our psychic knowledge can equal our physics. This could be done and can only be done by a specific training whereby the latent energy for this creative task and act, an energy which lurks inside us and manifests itself in pain and lust, individual neurosis and mass neurosis, be made express itself in advanced and purely psychological activities.

We can, then, have the three things we must gain or perish; we can

---

2. Many people still imagine with a European provincialism that with the Fall of the Roman Empire and the collapse of the already vitiated culture within its frontiers, the progress of civilization came to an end. We now know, however, that a frontier city such as Damascus kept going, through the Darkening Ages of the Decline and the Darkness of the Roman Fall, a balanced and advancing research in physics and psychology; metallurgy, ceramics, mysticism and explosives all being advanced, an advance which the Moslems were to permit and patronize; while research into the history of mathematics has shown how constant advances in this science (such as, for instance, the discovery of Zero and negative numbers), by which Europe benefited, were originated in India, the land of psychology. (Note: See Danzig's *History of Number.)*

answer rightly the three riddles which the sphinx demands of us on pain of destruction: (1) How can we reunite our sundered psyche and so cease to be neurotic? We can do this by knowing ourselves and when we know ourselves and are again one we shall have the full powers that belong to our united integrated selves. (2) How can we make our psychology balance our physics? By trained advance in increasing consciousness and enlarging our awareness. By this advance in psychological power we may at last achieve a non-violent sanction for all human relationships and a new approach in our relationships with animal life and a new insight into the inanimate. (3) How can we once again have direct union with Reality? We can do so by this advance in consciousness which resolves that strangulating of consciousness which we call the individual foreconscious. Having done this we know the meaning of Life, rewin our union with Nature and, finally, are able to reinterpret correctly the experience which we call Time and, doing so, we see Reality no longer distorted, but as it is.

Then we shall have understood our experience past, present, and to come, and acting on that understanding we shall have fulfilled the purpose of our Being, the meaning of evolution. We cannot say whether, with that discovery made and demonstrated, any but a fragment of mankind will attend to it or, attending, think the price, that discovery demands of each, if it is to be actual, should be paid. They may be able to resist seeing the vision and even when they see it the cost of entering into it through an altered life may seem prohibitive. We do know that that discovery can be made, and must be demonstrated and that those who perceive it and adhere to it find, not only the cost insignificant, but the life so disclosed to be a reality beside which the life left behind was a stifling shadow.